绿色"一带一路"
与2030年可持续发展议程
——有效对接与协同增效 共谋全球生态文明建设

周国梅　史育龙　[巴基斯坦]阿班·马克·卡布拉基（Aban Marker Kabraji）
周　军　等/著

中国环境出版集团·北京

图书在版编目（CIP）数据

绿色"一带一路"与2030年可持续发展议程：有效对接与协同增效 共谋全球生态文明建设：中国环境与发展国际合作委员会专题政策研究报告：汉文、英文/周国梅等著. —北京：中国环境出版集团，2021.7
ISBN 978-7-5111-4703-5

Ⅰ. ①绿… Ⅱ. ①周… Ⅲ. ①环境保护政策－研究报告－中国－汉、英 ②环境保护－可持续发展－研究报告－中国－汉、英 Ⅳ. ①X012②X22

中国版本图书馆CIP数据核字（2021）第069063号

出 版 人	武德凯
责任编辑	黄　颖
责任校对	任　丽
封面设计	岳　帅

出版发行 中国环境出版集团
（100062　北京市东城区广渠门内大街16号）
网　　址：http://www.cesp.com.cn
电子邮箱：bjgl@cesp.com.cn
联系电话：010-67112765（编辑管理部）
　　　　　010-67162011（第四分社）
发行热线：010-67125803，010-67113405（传真）

印　　刷	北京中科印刷有限公司
经　　销	各地新华书店
版　　次	2021年7月第1版
印　　次	2021年7月第1次印刷
开　　本	787×1092　1/16
印　　张	9.75
字　　数	150千字
定　　价	58.00元

【版权所有。未经许可，请勿翻印、转载，违者必究。】
如有缺页、破损、倒装等印装质量问题，请寄回本集团更换

中国环境出版集团郑重承诺：
中国环境出版集团合作的印刷单位、材料单位均具有中国环境标志产品认证；
中国环境出版集团所有图书"禁塑"。

专题政策研究项目组成员

中国环境与发展国际合作委员会

顾 问

汉　森	中国环境与发展国际合作委员会外方首席顾问， 加拿大可持续发展研究院高级顾问、前院长（加拿大）
郭　敬	生态环境部国际合作司司长
Thomas Lovejoy	联合国基金会高级研究员，乔治·曼森大学环境科学与政策教授

中外组长

周国梅	生态环境部对外合作与交流中心主任
Aban Marker Kabraji	世界自然保护联盟亚洲区办公室主任（巴基斯坦）
史育龙	中国国家发展和改革委员会城市和小城镇改革发展中心主任

中外成员

谷树忠	国务院发展研究中心资源与环境政策研究所副所长
张建平	商务部国际贸易经济合作研究院区域经济合作研究中心主任
葛察忠	生态环境部环境规划院环境政策部主任
孙轶颋	国际金融论坛副秘书长
陈　迎	中国社会科学院可持续发展研究中心副主任
王　苒	对外经济贸易大学全球价值链研究院绿色价值链研究室副主任
Ghiara Gianluca	欧洲—中国生态城市链接欧洲项目部核心专家，

	基伊埃可再生能源工程咨询公司法人代表及执行官（意大利）
Diana Mangalagiu	英国牛津大学环境变化学院教授， 法国诺欧商学院副教授（罗马尼亚、法国）

协调员

周　军	生态环境部对外合作与交流中心政策研究部副处长
朱春全	世界自然保护联盟中国代表处首席代表

支持专家

殷　红	中国工商银行城市金融研究所副所长， 中国金融学会绿色金融专业委员
崔　成	中国国家发展和改革委员会能源研究所处长
卢　伟	中国国家发展和改革委员会中国宏观经济研究院国土开发与地区经济研究所区域战略室副主任
蓝　艳	生态环境部对外合作与交流中心高级工程师
彭　宁	生态环境部对外合作与交流中心工程师
黄一彦	生态环境部对外合作与交流中心工程师
李盼文	生态环境部对外合作与交流中心工程师
赵海珊	生态环境部对外合作与交流中心工程师
于心怡	生态环境部对外合作与交流中心助理工程师
杜艳春	生态环境部环境规划院环境政策部工程师
何　琦	对外经济贸易大学全球价值链研究院绿色价值链研究室助理研究员
张　诚	世界自然保护联盟华南项目主任
杨　艳	国务院发展研究中心资源与环境政策研究所工程师
韩珠萍	商务部国际贸易经济合作研究院工程师
杜晶晶	商务部国际贸易经济合作研究院工程师

SPECIAL POLICY STUDY TEAM MEMBERS

China Council for International Cooperation on Environment and Development

Senior Advisors

Art Hanson	CCICED International Chief Advisor Senior Advisor & Former Director, Institute of Sustainable Development, Canada
GUO Jing	Director General, Department of International Cooperation, Ministry of Ecology and Environment of China (MEE)
Thomas Lovejoy	Senior Researcher, United Nations Foundations Professor, Environmental Science and Policy, George Mason University

Co-team Leaders

ZHOU Guomei	Director General, Foreign Environmental Cooperation Center, MEE
Aban Marker Kabraji	Regional Director, Asia Regional Office of IUCN (Pakistan)
SHI Yulong	Director, General, China Center for Urban Development, National Development and Reform Commission (NDRC)

Core Experts

GU Shuzhong	Deputy Director, Research Institute of Resources and Environment Policies, Development Research Center of the State Council (DRC)
ZHANG Jianping	Director, Regional Economic Cooperation Research Center, Chinese Academy of International Trade and Economic Cooperation, Ministry of Commerce of China (MOFCOM)
GE Chazhong	Director, Environmental Strategy Institute, Chinese Academy for Environmental Planning, MEE
SUN Yiting	Deputy Secretary General, International Finance Forum (IFF)

CHEN Ying	Deputy Director General, Research Centre for Sustainable Development (RCSD), Chinese Academy of Social Sciences (CASS)
WANG Ran	Deputy Chief of Research Institute for Global Value Chain (RIGVC), University of International Business and Economics (UIBE)
Ghiara Gianluca	Key Expert, European Program EC-Link (Europe China Eco Cities Link) Legal Representative & Executive Officer, Geapower- Renewable Energy Engineering Consulting Co Ltd. (Italy)
Diana Mangalagiu	Professor of Environmental Change Institute University of Oxford, Associate Professor of French NEOMA Business School (Romania, France)

International Coordinators

ZHOU Jun	Deputy Director, Division for Policy Research, Foreign Enviromental Cooperation Center, MEE
ZHU Chunquan	Head of the IUCN China Office

Support Team

YIN Hong	Deputy Director, Urban Finance Research Institute of Industrial and Commercial Bank of China (ICBC), Member of Green Finance Committee of China Society for Finance and Banking
CUI Cheng	Director, Energy Research Institute, National Development and Reform Commission (NDRC)
LU Wei	Deputy Section Chief, Division of Regional Strategy, Institute of Spatial Planning and Regional Economy, China Academy of Macroeconomics Research, NDRC
LAN Yan	Senior Engineer, Foreign Environmental Cooperation Center, MEE
PENG Ning	Deputy Section Chief, Foreign Environmental Cooperation Center, MEE
HUANG Yiyan	Engineer, Foreign Environmental Cooperation Center, MEE
LI Panwen	Engineer, Foreign Environmental Cooperation Center, MEE
ZHAO Haishan	Engineer, Foreign Environmental Cooperation Center, MEE
YU Xinyi	Assistant Engineer, Foreign Environmental Cooperation Center, MEE

DU Yanchun	Engineer, Environmental Strategy Institute, Chinese Academy for Environmental Planning, MEE
HE Qi	Assistant, Research Fellow Green Global Value Chains Studies, Research Institute for Global Value Chains, University of International Business and Economics
ZHANG Cheng	IUCN South China Programme Manager
YANG Yan	Engineer, Research Institute of Resources and Environment Policies, Development Research Center of the State Council (DRC)
HAN Zhuping	Engineer, Regional Economic Cooperation Research Center, Chinese Academy of International Trade and Economic Cooperation, MOFCOM
DU Jingjing	Engineer, Regional Economic Cooperation Research Center, Chinese Academy of International Trade and Economic Cooperation, MOFCOM

前　言

联合国2030年可持续发展议程的启动带领国际社会迈入了可持续发展的新阶段，建设绿色"一带一路"与中国生态文明建设的理念高度契合，也顺应全球可持续发展的总体趋势。"一带一路"与联合国2030年可持续发展议程在理念、原则和目标方面高度契合、相辅相成，其倡导的"政策沟通、设施联通、贸易畅通、资金融通、民心相通"与2030年可持续发展议程的17个目标高度对应，它作为推动落实可持续发展议程的解决方案之一已经被国际社会认可。协同推进绿色"一带一路"与2030年可持续发展议程，将为区域可持续发展提供重要路径，避免发展中国家重走"先污染，后治理"的发展道路，将"一带一路"打造成全球生态文明和人类绿色命运共同体的重要载体。

"一带一路"倡议自2013年秋季提出以来，秉持"共商、共建、共享"的理念，得到国际社会大多数成员的认可。"一带一路"已成为全球公共产品，不仅能够向世界提供基础设施互联互通这样物质上的公共产品，更能提供制度理念上的公共产品，这将形成更加公正合理的全球治理理念和治理体系，推动各国一道解决环境、气候、扶贫等全球问题。

本书以中国环境与发展国际合作委员会（以下简称国合会）绿色"一带一路"与2030年可持续发展议程专题政策研究第一期项目（2019年度）研究成果为基础编辑而成，"一带一路"绿色发展国际联盟及其合作伙伴参与了该项研究。研究旨在分析绿色"一带一路"建设对落实 2030 年可持续发展目标的潜在贡献，识别绿色"一带一路"建设中的机遇和挑战，"一带一路"绿色发展的总体原则、目标和实施路径。在此，特别感谢参与研究工作的中、外专家，他们分别是：

执行摘要：周国梅、周军、蓝艳；

第一章：周国梅、周军、葛察忠、陈迎、蓝艳、杜艳春；

第二章：Diana Mangalagiu、Ghiara Gianluca；

第三章：史育龙、卢伟、崔成、张建平、孙轶颋、王苒、何琦、韩珠萍、杜晶晶；

第四章：Aban Marker Kabraji、谷树忠、史育龙、葛察忠、卢伟、朱春全、张诚、杜艳春、杨艳；

第五章：蓝艳、周军、史育龙、卢伟、殷红。

还要特别感谢彭宁、黄一彦、李盼文、赵海珊、于心怡等为本书的翻译、整理和编辑做出的贡献。

Forward

The implementation of the United Nations *2030 Agenda for Sustainable Development* led the international community into a new era of pursuing sustainable development. Building Green Belt and Road meets the requirement of ecological civilization construction and responds to the trend of the times. The Belt and Road Initiative (BRI) and the *2030 Agenda for Sustainable Development* have shared concepts, principles and goals. Policy coordination, facilities connectivity, unimpeded trade, financial integration and people-to-people bond proposed by the Belt and Road Initiative correspond to the 17 goals proposed by the *2030 Agenda for Sustainable Development*. The Belt and Road Initiative has been recognized as an effective solution to promoting the implementation of the 2030 Agenda globally. The synergistic development of Green Belt and Road and the *2030 Agenda for Sustainable Development* will facilitate sustainable development in the region, help developing countries to get rid of the mindset of "developing first and taking green issues in a second stage", and making BRI a platform for ecological civilization construction and the building of a green community of shared destiny.

Since its initiation in the Autumn of 2013, the Belt and Road Initiative has been widely recognized and positively received in the international community. The Belt and Road Initiative has become a global public good, both in terms of infrastructure connectivity and systems and concepts, which will contribute to the formation of a fair and reasonable global governance system and promote global efforts in addressing shared challenge in environment, climate and poverty reduction.

This book is compiled on the basis of the research results of the China Council for International Cooperation on Environment and Development (referred to as the CCICED) 2019 Special Policy Study on Green BRI and the *2030 Agenda for Sustainable Development*. BRI International Green Development Coalition (BRIGC) and its partners participated in this research.The research aims to analyze the potential contribution of green BRI construction to the implementation of the 2030 Sustainable Development Goals, to identify opportunities and challenges during the process, to study the long-term mechanism, important concerns and green development cases to promote the green development of BRI, and thus put forward policy recommendations in this regard. Here, special thanks to the Chinese and international experts who participated in the Special Policy Study. They are:

Executive Summary: Zhou Guomei, Zhou Jun, Lan Yan;

Chapter 1: Zhou Guomei, Zhou Jun, Ge Chazhong, Chen Ying, Lan Yan, Du Yanchun;

Chapter 2: Diana Mangalagiu、Ghiara Gianluca;

Chapter 3: Shi Yulong, Lu Wei, Cui Cheng, Zhang Jianping, Sun Yiting, Wang Ran, He Qi, Han Zhuping, Du Jingjing;

Chapter 4: Aban Marker Kabraji, Gu Shuzhong, Shi Yulong, Ge Chazhong, Lu Wei, Zhu Chunquan, Zhang Cheng, Du Yanchun, Yang Yan;

Chapter 5: Lan Yan, Zhou Jun, Shi Yulong, Lu Wei, Yin Hong.

Meanwhile, we would also like to thank the following colleagues for their contribution on translation, organizing and editing of this book, including Peng Ning, Huang Yiyan, Li Panwen, Zhao Haishan, Yu Xinyi.

目录 Contents

执行摘要 /3

1 绿色"一带一路"与2030年可持续发展议程 /5

1.1 "一带一路"及其建设进展 /5
1.2 绿色"一带一路"及其推进现状 /6
1.3 建设绿色"一带一路",将绿色发展由机遇转变为现实 /21
1.4 绿色"一带一路"对落实2030年可持续发展目标的潜在贡献 /22

2 绿色"一带一路"建设中的机遇与挑战 /26

2.1 机遇 /26
2.2 挑战 /27

3 "一带一路"建设重要关注问题研究 /29

3.1 推动绿色"一带一路"建设形成长效机制 /29
3.2 "一带一路"建设中绿色金融的战略布局与实施机制 /31
3.3 "一带一路"与构建更具包容性的全球绿色价值链 /36

4 "一带一路"绿色发展案例研究 /39

4.1 巴基斯坦、斯里兰卡实地调研 /39
4.2 中国—马来西亚钦州产业园 /45

5 推动绿色"一带一路"建设政策建议 /49

5.1 积极参与全球环境治理与气候治理,将"一带一路"打造成全球生态文明和绿色命运共同体的重要载体 /49
5.2 建立"一带一路"建设对接机制,以政策、规划、标准和技术对接促进战略对接并落地 /50
5.3 构建绿色"一带一路"源头预防机制,以绿色金融、生态环境影响评价等机制引导绿色投资 /52
5.4 构建"一带一路"项目管理机制,推动企业落实绿色发展实践 /54
5.5 通过民心相通加强绿色"一带一路"建设,强化人员交流与能力建设 /55

Executive Summary /59

1 Green Belt and Road and 2030 Agenda for Sustainable Development /62

- 1.1 The proposal and development of the Belt and Road Initiative /62
- 1.2 Progress of building a green Belt and Road /64
- 1.3 Promoting the green development of the Belt and Road Initiative, transforming opportunities into reality /85
- 1.4 The potential contribution of Green Belt and Road to the implementation of the 2030 SDGs /87

2 Opportunities and Challenges in the Development of Green Belt and Road /92

- 2.1 Opportunities /92
- 2.2 Challenges /93

3 Major Issues on the Belt and Road /96

- 3.1 Long-term mechanism of constructing green Belt and Road /96
- 3.2 Strategic arrangement and implementation mechanism of green finance in the development of the Belt and Road Initiative /100
- 3.3 The Belt and Road and constructing a more inclusive global green value chain /108

4 Case Studies on Green Development on the Belt and Road /111
    4.1 Field studies on Pakistan and Sri Lanka /111
    4.2 China-Malaysia Qinzhou industrial park /120

5 Policy Suggestions on Promoting Green Belt and Road /127
    5.1 Play an active role in global environmental governance and climate governance, transforming the Belt and Road Initiative into an important instrument for global ecological civilization construction and building a green community of common destiny /128
    5.2 Promoting strategic alignment in the development of the green Belt and Road with connection of policies, planning, standards and technologies /130
    5.3 Safeguard mechanisms for constructing a green Belt and Road from its source, and guiding green investment with mechanisms of green finance and ecological impact assessment /132
    5.4 Constructing a mechanism for BRI project management and promoting the business to adopt practices on green development /136
    5.5 Building a green Belt and Road through enhancing people- to-people bond, and enhancing personnel exchange and capacity building /138

Part 1 中文部分

绿色"一带一路"与 2030 年可持续发展议程
——有效对接与协同增效 共谋全球生态文明建设

Green BRI and 2030 Agenda for Sustainable Development
——Accelerating Effective Alignment and Synergy between BRI and 2030 Agenda to Promote Global Ecological Civilization

执行摘要

2015 年联合国大会通过的 2030 年可持续发展议程及《巴黎协定》标志着国际社会迈入了可持续发展的新阶段。中国也始终致力于推进绿色丝绸之路建设。2017 年 5 月，在"一带一路"国际合作高峰论坛上，中国国家主席习近平提出"践行绿色发展的新理念，倡导绿色、低碳、循环、可持续的生产生活方式，加强生态环保合作，建设生态文明，共同实现 2030 年可持续发展目标"。2017 年 4 月，环境保护部（现生态环境部）、外交部、国家发展改革委、商务部联合印发了《关于推进绿色"一带一路"建设的指导意见》，2017 年 5 月，环境保护部发布了《"一带一路"生态环境保护合作规划》，两大文件明确了未来 10 年"一带一路"生态环保工作的总体方向和全方位合作格局。

建设绿色丝绸之路顺应了国际社会绿色发展的潮流与趋势，也适应了发展中国家要求绿色发展、保护环境的现实需求，与 2030 年可持续发展议程的理念高度契合。总体来看，"一带一路"沿线多为发展中国家和新兴经济体，普遍面临着工业化、城市化带来的发展与保护的矛盾。绿色"一带一路"建设将有助于推动沿线国家开展环境合作，促进沿线国家环境管理能力提升，助力区域落实 2030 年可持续发展目标。

"一带一路"在 6 年多的实践中，始终秉持绿色发展理念，注重与联合国 2030 年可持续发展议程的对接，推动基础设施绿色低碳化建设和运营管理，在投资贸易中强调生态文明理念，加强生态环境、生物多样性保护和应对气候变化领域的合作，为落实 2030 年可持续发展目标提供了新的动力，也给相关国家绿色发展带来了新机遇。一是绿色"一带一路"将推动生态环保政策贯通，在政策层面对接绿色"一带一路"与 2030 年可持续发展议程；二是防范设施联通环境风险，打造一体化的生态环境风险防范和治理体系；三是促进绿色贸易畅通，推动可持续生产与消费，加强绿色供应链管理；四是推动绿色资金融通，引导资金向清洁能源等领域的投资；五是加强环保民心相通，推动发展中国家生态环保能力建设。

"一带一路"是一项宏大的工程，在创造大量机遇的同时，也面临着一系列挑战。在绿色理念方面，部分共建"一带一路"国家发展水平还不高，生态环境保护这一理念还未得到充分重视；在监控评估方面，"一带一路"项目大多涉及多个国家，在项目规划、设计、建设、运营和评估的过程中会面对不同的标准和流程；在绿色金融与投资方面，还缺乏一定的政策引导。从项目来看，"一带一路"项目多为大型基础设施项目，在为当地经济发展、就业民生等创造机遇的同时，也会面临一定的环境和社会风险。

"一带一路"是绿色发展之路，需要各方共同努力，积极作为。建设绿色丝绸之路，就是要践行"绿水青山就是金山银山"的理念，共谋全球生态文明建设之路。建设绿色"一带一路"将为共建国家和地区创造更多的绿色公共产品，有效推动2030年可持续发展议程的落实。我们相信，在国内外合作伙伴的携手努力下，绿色"一带一路"建设将会取得丰硕成果。

研究认为，共建绿色"一带一路"需要凝聚各方共识，从战略、规划、评估和金融支持等层面，探索建立绿色"一带一路"建设的长效机制。而建立完善的绿色"一带一路"长效机制要从战略对接、规划嵌入、项目评估、金融支持、技术推广等方面入手。此外，构建更具包容性的全球绿色价值链，也是打造绿色"一带一路"的关键抓手。

本书研究分析了巴基斯坦、斯里兰卡"一带一路"项目生态环保情况，以及中国—马来西亚钦州产业园等绿色发展案例，为"一带一路"项目，尤其是基础设施建设项目提出了绿色发展原则，从具体实施层面明确了"一带一路"与2030年可持续发展议程的对接途径。

基于扎实的理论基础和实践总结，本书提出了推动绿色"一带一路"建设的政策建议：一是积极参与全球环境治理与气候治理，将"一带一路"打造成全球生态文明和绿色命运共同体的重要载体；二是建立"一带一路"倡议对接机制，以政策、规划、标准和技术对接促进战略对接并落地；三是构建绿色"一带一路"源头预防机制，以绿色金融、生态环境影响评价等机制引导绿色投资；四是构建"一带一路"项目管理机制，推动企业开展绿色发展实践活动；五是通过民心相通加强绿色"一带一路"建设，强化人员交流与能力建设。

1 绿色"一带一路"与2030年可持续发展议程

1.1 "一带一路"及其建设进展

自2008年全球金融危机爆发以来,世界经济持续低迷,贸易增长缓慢,各种不稳定事件频出。世界经济迫切需要新的增长动力并建立新的循环。发展中国家包括新兴经济体对基础设施以及产业发展的巨大需求有望成为世界经济新的增长动力。

在此背景下,2013年,中国国家主席习近平在哈萨克斯坦和印度尼西亚提出共建"丝绸之路经济带"和"21世纪海上丝绸之路",即"一带一路"倡议。2015年3月,中国政府发布了《推动共建丝绸之路经济带和21世纪海上丝绸之路的愿景与行动》[①](以下简称《愿景与行动》),从目标愿景、合作原则、重点方向等方面提出了共建"一带一路"的顶层设计框架。根据《愿景与行动》,"一带一路"旨在促进经济要素有序自由流动、资源高效配置和市场深度融合,推动沿线各国实现经济政策协调,开展更大范围、更高水平、更深层次的区域合作,共同打造开放、包容、均衡、普惠的区域经济合作架构。

"共商、共建、共享"是"一带一路"建设最根本的原则,其核心是倡导共建"一带一路"国家发展战略对接,最大限度地凝聚各方共识,发挥各自的优势共同参与建设,共同分享"一带一路"建设项目的成果以及长期发展红利。"一带一路"建设以政策沟通、设施联通、贸易畅通、资金融通、民心相通为合作重点。

"一带一路"倡议自提出以来,得到共建国家和国际社会的积极响应,共建"一带一路"正在成为中国参与全球开放合作、改善全球经济治理体系、促进全球共同发展繁荣、推动构建人类命运共同体的中国方案。2017年5月,首届"一带一路"国际合作高峰论坛顺利召开,来自29个国家的元首和政府首脑、130多个国

---

① 中国国家发展改革委,外交部,商务部. 推动共建丝绸之路经济带和21世纪海上丝绸之路的愿景与行动 [EB/OL]. 2015-03-28. http://www.mofcom.gov.cn/article/resume/n/201504/20150400929655.shtml.

家和70多个国际组织的代表,再次确认了"共商、共建、共享"的核心理念。2019年4月,第二届"一带一路"国际合作高峰论坛成功举办,包括中国在内的38个国家的元首和政府首脑等领导人以及联合国秘书长和国际货币基金组织总裁共40位领导人出席圆桌峰会。来自150个国家、92个国际组织的6 000余名外宾参加了论坛。与会各方就共建"一带一路"深入交换意见,普遍认为"一带一路"是机遇之路,就高质量共建"一带一路"达成了广泛共识,取得了丰硕成果。

经过6年的不懈努力,"一带一路"建设从理念、愿景转化为现实行动,已进入落地生根、开花结果的全面推进阶段。2013—2018年,中国与沿线国家货物贸易进出口总额超过6万亿美元。在沿线国家建设的境外经贸合作区总投资达200多亿美元,创造了就业岗位数十万个,给当地创造的税收达几十亿美元。一批合作项目取得实质性进展,中巴经济走廊建设进展顺利,中老铁路、中泰铁路、匈塞铁路建设稳步推进,雅万高铁部分路段已经开工建设,瓜达尔港已具备全作业能力。截至2019年年底,中欧班列累计开行数量突破2.1万列,已经联通亚欧大陆17个国家的115个城市。

1.2 绿色"一带一路"及其推进现状

1.2.1 "一带一路"重点地区生态环境现状

共建"一带一路"国家环境和气候差异较大,生态环境问题突出。东南亚、南亚、西亚、北非等区域大部分国家都是发展中国家,随着人口的过快增长和工业化的快速发展,资源消耗和污染物随意排放令环境状况不容乐观,森林面积缩小、水污染和空气污染日趋严重,环境污染趋势未能得到有效遏制。

(1)生态环境总体敏感

总体来看,共建"一带一路"国家面临着各种生态环境问题,生态环境总体较为敏感。

中巴经济走廊新疆段干旱缺水。巴基斯坦北部地区位于喀喇昆仑、喜马拉雅、兴都库什等高山山脉交汇处,生物多样性丰富,为走廊互联互通带来巨大挑战。巴基斯坦西南部地区干旱缺水问题严重,而东南部地区存在大气、水、土壤污染以及季节性缺水等问题。

中蒙俄经济走廊的俄罗斯西南部地区，地处寒带亚寒带，永久性冻土带面积较广，森林砍伐较严重，生态系统一旦被破坏，将很难恢复。蒙古高原区地表水资源匮乏，植被覆盖率低，荒漠生态环境脆弱。

中国—中南半岛经济走廊的柬埔寨、泰国、老挝、越南等湄公河流域国家水资源丰富，森林覆盖率较高，植物种类多样，水资源争端和水环境污染问题比较突出。马来半岛平原区面积狭小，存在一定的污染问题。

中国—中亚—西亚经济走廊主体为高山—绿洲—干旱荒漠，水资源短缺和跨界水资源争端明显。中亚国家存在土地荒漠化、遗留核污染等问题。

孟中印缅经济走廊的高山高原区森林近年来遭到不同程度的破坏，导致生态景观破碎化和生物多样性减少。低山平原区的农业面源污染和耕地退化等问题突出。孟加拉湾沿海地区存在沿海湿地红树林破坏以及气候变化带来的海平面上升等问题。

新亚欧大陆桥经济走廊西部地区干旱和荒漠化明显。中亚段干旱和脆弱的荒漠生态系统是其生态环境面临的主要挑战，而俄罗斯段与欧洲段沿线的各类自然保护区较多。

（2）地区性水污染问题严重影响区域的经济发展和社会稳定

中亚地区地处欧亚大陆腹地，以沙漠和草原地形为主，气候干燥，水资源短缺和水环境污染是该地区最主要的环境问题。具体表现在以下几个方面：一是湖泊面积不断缩小，且水质下降；二是河流水量减少，河流缩短或消失，且水质下降；三是地下水位下降，水质变差；四是盐碱化土地面积增加；五是沙漠扩大，绿洲缩小，沙尘暴频度上升；六是自然植被减少，植被类型退化。这些变化在咸海流域表现最为明显，由于水质恶化，该地区人口出生率下降，婴儿死亡率上升，很多居民迫于生态环境压力而迁居。水资源成为阻碍中亚社会经济发展的主要因素之一。

南亚地区有不少国家也面临严重的水污染问题。例如，印度很多城市，大多数水体正在遭受生活污水、工业废水、化学药品和固体废物的严重污染。水污染、洪灾和旱灾已成为印度面临的与水有关的三大"灾害"，已经影响了印度社会的可持续发展。由于在水资源保护方面投入不足，印度每天有大量的工业废水直接排入河流、湖泊及地下，造成地下水大面积污染，一些化学物质严重超标，如铅含量比废水处理较好的工业化国家高 20 倍。此外，直接排放未经处理的生活污水也加剧了水污染程度，严重影响着百姓健康。流经印度北方的主要河流——恒河已被列为世界上污染最严重的河流之一。当地居民饮用和烹饪时使用受污染的地下水已经导致了许多健康问题。同时，由于地下水污染严重，目前在印度市场上销售的 12 种软饮料的有害残留物含量均超标。有些软饮料中杀虫剂残留物含量超过欧洲标准 10~70 倍。

中东地区干旱少雨，水资源紧缺，形成浩瀚的沙漠，人口大多集中在沿海或有限的江河流域地区。近 20 年来，大规模的工业化以及相对滞后的经济社会发展水平，又使得中东地区生态环境进一步恶化，特别是在石油生产国，环境问题尤为突出。包括硫酸盐、硝酸盐在内的各种污染物，使得科威特、卡塔尔、沙特阿拉伯、伊朗、伊拉克等石油生产国的生态环境受到明显损害，也对周围其他中东国家的生态环境形成一定威胁。20 世纪 80 年代的两伊战争和 90 年代初的海湾战争更是给海湾地区带来了严重的"环境后遗症"：石油设施和油轮在战争中被破坏，导致大量石油泄漏，流入海湾水域，造成大面积污染，直接危及该地区海洋鱼类的生存。

（3）空气污染问题在"一带一路"沿线国家也比较突出

"一带一路"沿线国家的空气污染水平整体低于全球平均水平，但个别国家和地区却存在比较严重的空气污染问题。

南亚、西亚和北非地区饱受细颗粒物（$PM_{2.5}$）污染困扰。"一带一路"沿线 65 个国家中有 22 个国家 $PM_{2.5}$ 浓度较高，其中有 11 个国家位于西亚、北非地区，6 个国家位于南亚地区。印度德里、巴基斯坦拉合尔、孟加拉国达卡等城市是全球空气污染最严重的几个城市。除受制于地理条件——空气扩散条件较差以外，这两个地区过分依赖石油化工和重化工业的发展模式也是导致颗粒物浓度偏高的重要原因。

东北亚、东南亚地区的空气污染形势也比较严峻。以蒙古国首都乌兰巴托为例，快速的城市化和工业化发展导致城市空气质量迅速恶化，另外，家用散煤、机动车尾气、工业和建筑活动的电力需求等都加大了空气污染物排放。世界卫生组织 2011 年空气质量排名数据中，将乌兰巴托市列为世界上空气质量最差的前五个城市之一。东南亚地区则是由于森林大火、工业污染和汽车尾气等因素导致空气质量下降,越南河内、泰国曼谷、印度尼西亚等地最为严重。除此之外,东欧地区由于重工业发展和发电厂建设也面临着严重的空气污染问题。

1.2.2 "一带一路"沿线国家可持续发展目标指数分布

(1) 环境目标是2030年可持续发展目标(SDGs)的重要构成

从20世纪60年代起,国际社会就开始反思和探寻经济、社会与环境协调发展之路。2000年,在联合国千年首脑会议上,189个国家正式签署《联合国千年宣言》,并庄重做出承诺,就全世界范围内消除贫困、饥饿、文盲、性别歧视,减少疾病传播,阻止环境恶化,商定了一套到2015年达成的目标,即联合国千年发展目标(Millennium Development Goals,MDGs)。在千年发展目标的指导下,全球在发展经济、改善人居、消除饥饿与贫困等方面取得了积极的进展。考虑MDGs于2015年完成历史使命,2012年6月在巴西里约热内卢召开的联合国可持续发展大会(即"里约+20"峰会),将制定一套以千年发展目标为基础、以可持续发展为核心的2015后全球发展议程提上日程。2015年9月25日,联合国可持续发展峰会正式通过了《变革我们的世界:2030年可持续发展议程》(以下简称《2030年议程》),建立了全球可持续发展目标(Sustainable Development Goals,SDGs),为未来15年世界各国的发展和国际合作指明了方向,勾画了蓝图。

可持续发展目标(SDGs)是《2030年议程》的核心内容之一。《2030年议程》共分四大部分:政治宣言、全球可持续发展目标、执行手段、后续行动①。《2030年议程》不再局限于发展中国家,而是适用于世界各国。可持续发展目标(SDGs)是《2030年议程》的核心内容,涵盖了经济、社会和资源环境三大领域,包括了17个可持续发展目标和169个子目标。这些目标的实施期限是到2020年或2030年,旨在根本性改变片面追求经济增长的传统发展观,指导全球各国在今后15年内的发展政策和资金使用,在人类和地球至关重要的领域采取行动,消除贫困、保护地球、确保所有人共享繁荣。其中,前16个目标是国际社会致力于实现的目标,第17个目标(全球发展伙伴关系)则强

---

① 中华人民共和国外交部. 变革我们的世界:2030年可持续发展议程[EB/OL]. 2016. http://infogate.fmprc.gov.cn/ web/ziliao_674904/zt_674979/dnzt_674981/qtzt/2030kcxfzyc_686343/t1331382.shtml.

调通过国际社会携手合作来推动《2030年议程》的落实。

环境目标是可持续发展目标的重要组成部分。《2030年议程》强调资源、环境带来的生存、生活方面的挑战，环境目标几乎直接或间接体现在可持续发展目标的所有目标与指标中，涉及生态环境保护的各个方面。通过对可持续发展目标和指标进行梳理发现，约有52.9%的总体目标和14.2%的子目标与生态环境保护相关，有些环境目标和指标是独立的，有些则将环境目标融入其他发展目标和指标中。其中，总目标涉及水环境安全（SDG6）、可持续能源（SDG7）、可持续工业化（SDG9）、可持续城市（SDG11）、可持续消费和生产（SDG12）、气候变化（SDG13）、海洋环境（SDG14）、陆地生态系统及生物多样性（SDG15）、可持续全球伙伴关系（SDG17）等9个方面（表1-1），其他总目标中涉及的环境子目标包括化学品污染防治、空气质量、土壤污染状况改善等方面。

表 1-1　可持续发展目标中的环境相关目标

总体目标		所涉及环境问题	子目标数量
目标6（SDG6）	为所有人提供水和环境卫生并对其进行可持续管理	水和环境卫生	8
目标7（SDG7）	确保人人获得负担得起的、可靠和可持续的现代能源	可持续现代能源	5
目标9（SDG9）	建造具备抵御灾害能力的基础设施，促进具有包容性的可持续工业化，推动创新	可持续工业化	8
目标11（SDG11）	建设包容、安全、有抵御灾害能力和可持续的城市和人类住区	可持续城市	10
目标12（SDG12）	采用可持续的消费和生产模式	可持续消费和生产	11
目标13（SDG13）	采取紧急行动应对气候变化及其影响	气候变化	5
目标14（SDG14）	保护和可持续利用海洋和海洋资源以促进可持续发展	海洋和海洋资源	10
目标15（SDG15）	保护、恢复和促进可持续利用陆地生态系统，可持续管理森林，防治荒漠化，制止和扭转土地退化，遏制生物多样性的丧失	陆地生态系统、森林、荒漠化、土地退化、生物多样性	12
目标17（SDG17）	加强执行手段，重振可持续发展全球伙伴关系	可持续全球伙伴关系	19

（2）2018 年可持续发展目标指数和指示板全球报告环境目标与指标

可持续发展目标指数（SDG Index）和指示板（SDG Dashboards）提供的是国别层面 SDGs 进展的测量方法。对《2030 年议程》中各个目标的度量和监测是执行该议程最重要的环节之一，因此建立一套有效的 SDGs 指标体系显得尤为必要。基于此，联合国可持续发展解决方案网络（SDSN）等机构提出了可持续发展目标指数和指示板，它不是官方监测工具，而是通过联合国官方 SDGs 指标和其他可靠数据弥补相关数据缺失，在此基础上利用经济合作与发展组织（OECD）（2008）国家构建综合指标的方法，提出关键性假设，制定出的一套用于国家层面的测量标准。这套指数也是在联合国统计司的支持下，由各成员国发起的对 SDGs 官方指标的补充和支持，旨在帮助各个国家解决在实现 SDGs 的过程中筛选出的优先问题，明确挑战和差距，以促进实现更加有效的可持续发展决策。

2015—2018 年，SDSN 与贝塔斯曼基金会（Bertelsmann Foundation）联合发布了可持续发展目标指数和指示板全球报告。2015 年，SDSN 与贝塔斯曼基金会发布了名为《可持续发展目标：富裕国家是否准备好了？》的报告，描述了 34 个 OECD 国家在可持续发展目标方面的实施现状。自 2016 年起，每年由 SDSN 与贝塔斯曼基金会联合发布可持续发展目标指数和指示板全球报告。2017 年 7 月，SDSN 与贝塔斯曼基金会联合发布了《2017 年可持续发展目标指数和指示板报告——全球责任：实现目标的国际溢出效应》。2018 年 7 月，SDSN 与贝塔斯曼基金会联合发布了《2018 年可持续发展目标指数和指示板报告——全球责任：实现目标》。报告总结了目前各国在 17 个可持续发展目标方面的表现和趋势，以可持续发展目标指示板展示，通过跨国家比较，有助于确定各国在实现可持续发展目标方面的成就和差距。该系列报告利用可持续发展目标指数对各国落实 17 个可持续发展目标的现状进行排名，并通过颜色编码体现 17 个总目标整体实施情况，最终以可持续发展目标指示板展示，并为每个国家的 SDGs 实施现状出具一份详细报告，为横向比较不同国家间的可持续发展水平提供了可能。

《2018 年可持续发展目标指数和指示板报告——全球责任：实现目标》以各类公开数据库为基础，对指标和方法做了更新，分析了全球 156 个国家实现可持续发展目标的优势与不足，为比较不同国家间的可持续发展水平提供了依据。总体来看,在 2018 年评估的可持续发展总目标和所有指标中,有 9 项总目标和 31 个指标与生态环境保护直接或间接相关(表 1-2),与 2017 年 29 个指标比较,增加了 2 个。具体来看,清洁能源(SDG7)在原有基础上加入获得清洁燃料和烹饪技术(人口百分比)指标;可持续消费与生产(SDG12)中增加城市固体废物产生量[kg/(a·人)]指标;应对气候变化(SDG13)中增加化石燃料出口中隐含的 CO_2 排放量(kg/人)指标。同时水和环境卫生的可持续管理(SDG6)剔除优质水源获得率(%)这一指标。

表 1-2 2018 年 SDG 指数和指示板报告与生态环境保护相关的 9 项总目标与 31 个指标

SDG	描述/标签
2	可持续氮管理指数
3	室内空气污染和环境空气污染导致的死亡率(每 10 万)
6	可再生性淡水资源获取量占总可再生性水资源量的比例(%)
6	国际农产品贸易中的进口商品隐含的地下水消耗量[m^3/(a·人)]
7	最终能源消费中的可再生能源的比例(%)
7	获得清洁燃料和烹饪技术(人口百分比)
7	燃料燃烧/发电产生的 CO_2 排放量[tCO_2/(亿 kW·h)]
11	城市地区 $PM_{2.5}$ 年平均质量浓度($\mu g/m^3$)
11	能获得管网中的安全可靠饮用水的城市人口比例(城市人口百分比)
12	电子垃圾产生量(kg/人)
12	城市固体废物产生量[kg/(a·人)]
12	废水处理率(%)
12	生产活动产生的 SO_2 排放量(kg/人)
12	国际贸易中进口商品隐含的 SO_2 净排放量(kg/人)
12	生产活动产生的氮足迹(kg/人)
12	进口商品隐含的活性氮净排放量(kg/人)
13	能源相关 CO_2 排放量(tCO_2/人)
13	技术改进条件下的进口商品隐含的 CO_2 排放量(tCO_2/人)
13	气候变化脆弱性指数(0~1)
13	化石燃料出口中隐含的 CO_2 排放量(kg/人)
13	有效碳汇率(欧元/tCO_2)

SDG	描述/标签
14	海洋关键生物多样性区域中受到妥善保护的平均比例（%）
	海洋健康指数 生物多样性（0~100）
	海洋健康指数 清洁水域（0~100）
	海洋健康指数 渔业（0~100）
	经济海域中鱼类资源过度开发或崩溃的比例（%）
15	陆地生物多样性关键区域中受到妥善保护的平均比例（%）
	淡水生物多样性关键区域中受到妥善保护的平均比例（%）
	红色名录物种生存指数（0~1）
	森林面积年度变化率（%）
	受国际贸易威胁的物种数量（物种/100万人）

（3）2018年"一带一路"沿线国家SDG指数整体情况

《2018年可持续发展目标指数和指示板报告——全球责任：实现目标》将国家范围由157个减少到156个。将"一带一路"沿线65个国家[①]与156个国家进行比对发现，除文莱、马尔代夫、吉尔吉斯斯坦、巴勒斯坦等4个国家外，有61个"一带一路"沿线国家在2018年SDG指数的核算范围内。表1-3列出了61个国家在实现17个可持续发展目标方面的指数得分及在全球156个国家中的排名。可以发现，东南亚、南亚、中亚、西亚、中东欧、东欧、北非等不同国家及地区SDG指数表现具有较大的差异性。

中东欧和东欧地区国家SDG指数排名靠前。报告共涉及61个"一带一路"沿线国家，其中包含中东欧国家16个、东欧国家4个，共计20个。总体来看，除马其顿、阿尔巴尼亚、黑山、波黑和俄罗斯外，其余15个国家位列61个沿线国家排名的前15名，较东南亚、南亚、中亚、西亚等区域有突出的优势。其中，斯洛文尼亚（第8位）、捷克（第13位）、爱沙尼亚（第16位）、克罗地亚（第21位）、白俄罗斯（第23位）在全球156个国家中排名靠前，距离实现2030年可持续发展目标最为接近。

---

① 王义桅. 世界是通的——"一带一路"的逻辑[M]. 北京：商务印书馆，2016：106.

表 1-3　2018 年 61 个"一带一路"沿线国家 SDG 指数及在全球 156 个国家中的排名

序号	国家	分值	全球156个国家中的排名	序号	国家	分值	全球156个国家中的排名
1	斯洛文尼亚共和国	80	8	25	亚美尼亚	69.3	58
2	捷克共和国	78.7	13	26	泰国	69.2	59
3	爱沙尼亚	78.3	16	27	阿拉伯联合酋长国	69.2	60
4	克罗地亚	76.5	21	28	马其顿	69	61
5	白俄罗斯	76	23	29	俄罗斯联邦	68.9	63
6	斯洛伐克共和国	75.6	24	30	阿尔巴尼亚	68.9	62
7	匈牙利	75	26	31	哈萨克斯坦	68.1	65
8	拉脱维亚	74.7	27	32	土耳其	66	79
9	摩尔多瓦	74.5	28	33	黑山共和国	67.6	69
10	波兰	73.7	32	34	波黑	67.3	71
11	保加利亚	73.1	34	35	塔吉克斯坦	67.2	73
12	立陶宛	72.9	36	36	巴林	65.9	80
13	乌克兰	72.3	39	37	伊朗伊斯兰共和国	65.5	82
14	塞尔维亚	72.1	40	38	不丹	65.4	83
15	以色列	71.8	41	39	菲律宾	65	85
16	新加坡	71.3	43	40	黎巴嫩	64.8	87
17	罗马尼亚	71.2	44	41	斯里兰卡	64.6	89
18	阿塞拜疆	70.8	45	42	约旦	64.4	91
19	格鲁吉亚	70.7	47	43	阿曼	63.9	94
20	塞浦路斯	70.4	50	44	蒙古国	63.9	95
21	乌兹别克斯坦	70.3	52	45	阿拉伯埃及共和国	63.5	97
22	中国	70.1	54	46	沙特阿拉伯	62.9	98
23	马来西亚	70	55	47	印度尼西亚	62.8	99
24	越南	69.7	57	48	尼泊尔	62.8	102

序号	国家	分值	全球156个国家中的排名	序号	国家	分值	全球156个国家中的排名
49	科威特	61.1	105	56	缅甸	59	113
50	卡塔尔	60.8	106	57	阿拉伯叙利亚共和国	55	124
51	老挝人民民主共和国	60.6	108	58	巴基斯坦	54.9	126
52	柬埔寨	60.4	109	59	伊拉克共和国	53.7	127
53	土库曼斯坦	59.5	110	60	阿富汗	46.2	151
54	孟加拉国	59.3	111	61	也门共和国	45.7	152
55	印度	59.1	112				

东南亚和中亚地区国家SDG指数排名差距较大。东南亚地区9个国家中，新加坡（第43位）、马来西亚（第55位）、越南（第57位）和泰国（第59位）在区域内领先，菲律宾（第85位）、印度尼西亚（第99位）、老挝（第108位）、柬埔寨（第109位）和缅甸（第113位）排名靠后。中亚地区4个国家中，乌兹别克斯坦（第52位）、哈萨克斯坦（第65位）排名靠前，在区域内领先，塔吉克斯坦（第73位）、土库曼斯坦（第110位）排名靠后。

西亚地区国家SDG指数排名比较分散。西亚地区19个国家中，以色列（第41位）、阿塞拜疆（第45位）、格鲁吉亚（第47位）、塞浦路斯（第50位）等4个国家的排名位于156个国家的前1/3；亚美尼亚（第58位）、阿拉伯联合酋长国（第60位）、土耳其（第79位）、巴林（第80位）、伊朗（第82位）、黎巴嫩（第87位）、约旦（第91位）、阿曼（第94位）、沙特阿拉伯（第98位）等9个国家的排名位于156个国家的中间1/3；科威特（第105位）、卡塔尔（第106位）、叙利亚（第124位）、伊拉克（第127位）、阿富汗（第151位）、也门（第152位）等6个国家排名靠后，位于156个国家的后1/3。

南亚国家SDG指数排名在后50%。南亚地区6个国家中，不丹（第83位）、斯里兰卡（第89位）、尼泊尔（第102位）、孟加拉国（第111位）、印度（第112位）、巴基斯坦（第126位）排名位于156个国家的后50%，实现2030年可持续发展目标面临的挑战较大。

1.2.3 绿色"一带一路"建设进展

绿色是"一带一路"倡议的重要内容。《愿景与行动》中提出"在投资贸易中突出生态文明理念,加强生态环境、生物多样性和应对气候变化合作,共建绿色丝绸之路"。2016 年 6 月,中国国家主席习近平在乌兹别克斯坦最高会议立法院发表演讲时指出,要践行绿色发展理念,携手打造绿色丝绸之路。2017 年 5 月,在首届"一带一路"国际合作高峰论坛上,习近平主席强调"践行绿色发展的新理念,倡导绿色、低碳、循环、可持续的生产生活方式,加强生态环保合作,建设生态文明,共同实现 2030 年可持续发展目标"。

2019 年 4 月,习近平主席在第二届"一带一路"国际合作高峰论坛开幕式上指出,我们要"把绿色作为底色,推动绿色基础设施建设、绿色投资、绿色金融,保护好我们赖以生存的共同家园","我们同各方共建'一带一路'可持续城市联盟、绿色发展国际联盟,制定《'一带一路'绿色投资原则》,发起'关爱儿童、共享发展,促进可持续发展目标实现'合作倡议。我们启动共建'一带一路'生态环保大数据服务平台,将继续实施绿色丝路使者计划,并同有关国家一道,实施'一带一路'应对气候变化南南合作计划"。

绿色"一带一路"的内涵是以生态文明与绿色发展理念为指导,坚持资源节约和环境友好原则,将绿色发展和生态环保融入"一带一路"建设的各方面和全过程。一是将其作为切入点,增进与共建国家的政策沟通;二是防控生态环境风险,保障与共建国家的设施联通;三是提高产能合作的绿色化水平,促进与共建国家的贸易畅通;四是完善投融资机制,服务与共建国家的资金融通;五是加强生态环保国际合作与交流,促进与共建国家的民心相通。从以上五个方面着手,为共建国家实现可持续发展目标中的环境指标做出直接贡献[①]。

"一带一路"作为带动共建国家经济发展的重大倡议,已经被国际社会认可为推动落实 2030 年可持续发展议程的解决方案之一。第 72 届联合国大会主席米罗斯拉夫·莱恰克表示,中国正在通过"一带一路"倡议分享财富和最佳实践,从而促进可持续发展目标的落实。联合国秘书长安东尼奥·古特

---

① 董战峰,葛察忠,王金南,等. "一带一路"绿色发展的战略实施框架[J]. 中国环境管理,2016,8(2):31-35,41.

雷斯指出，2030年可持续发展议程与"一带一路"倡议有共同的宏伟目标，都旨在创造机会，带来有益全球的公共产品，并在多方面促进全球互联互通，包括基础设施建设、贸易、金融、政策以及文化交流，从而带来新的市场和机会，"一带一路"倡议对议程的实施具有巨大的推动作用[①]。美国环保协会总裁弗莱德·克鲁伯也认为，"一带一路"倡议能够带来经济繁荣和环境改善的双重效益。

绿色"一带一路"的总体目标和任务措施进一步明确。 2017年4月，环境保护部（现生态环境部）、外交部、国家发展改革委、商务部联合印发了《关于推进绿色"一带一路"建设的指导意见》。2017年5月，环境保护部（现生态环境部）发布了《"一带一路"生态环境保护合作规划》，提出力争用3～5年时间建成务实高效的生态环保交流合作体系、支撑与服务平台和产业技术合作基地，制定落实一系列生态环境风险防范的政策和措施，用5～10年时间建成较为完善的生态环保服务、支撑、保障体系。该规划明确了将绿色发展融入"五通"的主要活动，并列出25个重点项目[②]。

搭建"一带一路"绿色发展国际伙伴合作平台。 为增进共建国家对绿色"一带一路"的了解和认知，中国同国际组织、共建"一带一路"国家、非政府组织积极开展绿色"一带一路"研讨、交流和对接。中国生态环境部和中外合作伙伴共同发起了"一带一路"绿色发展国际联盟，旨在搭建一个国际平台，分享绿色发展理念、政策与实践，开展研讨及对话，提升"一带一路"沿线区域实现环境治理目标的能力。通过举办中国—阿拉伯国家环境合作论坛、中国—东盟环境合作论坛、上海合作组织环保合作·中国活动周、澜沧江—湄公河环境合作周等活动，与共建"一带一路"国家积极开展政策对话。

---

① 联合国秘书长古特雷斯："一带一路"为应对全球性挑战提供新机遇[N]. 人民日报，2017-05-12.
② 程翠云，翁智雄，葛察忠，等. 绿色丝绸之路建设思路与重点任务——《"一带一路"生态环保合作规划》解读[J]. 环境保护，2017，45（18）：53-56.

"一带一路"绿色发展国际联盟

2017年5月,中国国家主席习近平在首届"一带一路"国际合作高峰论坛(以下简称高峰论坛)开幕式演讲中倡议,建立"一带一路"绿色发展国际联盟(以下简称联盟)。联盟由中外合作伙伴于2019年4月在第二届"一带一路"国际合作高峰论坛绿色之路分论坛上共同启动。联盟定位为一个开放、包容、自愿的国际合作网络,旨在推动将绿色发展理念融入"一带一路"建设,进一步凝聚国际共识,促进共建"一带一路"国家落实2030年可持续发展议程。截至2019年年底,已有130多家中外合作伙伴加入联盟,其中外方合作伙伴70多家,包括26个共建国家环境主管部门、国际组织、研究机构和企业等。

联盟旨在打造政策对话和沟通平台,分享生态文明和绿色发展的理念与实践,加强沟通交流,推动建设联合研究网络;打造环境知识和信息平台,推动建立生态环保信息共享机制,为绿色丝绸之路建设提供数据及分析支撑,并推动区域环境管理能力建设;打造绿色技术交流与转让平台,促进绿色低碳技术交流与转让,提高区域基础设施建设及相关投资贸易活动的绿色化水平。

联盟主要通过专题伙伴关系开展活动,涉及领域包括但不限于生物多样性和生态系统、绿色能源和能源效率、绿色金融与投资、环境质量改善和绿色城市、南南合作和可持续发展目标、绿色技术创新和企业社会责任、可持续交通、全球气候变化治理与绿色转型、环境法律法规和标准、海洋命运共同体和海洋环境治理等,现已启动10个专题。此外,联盟框架下还将开展"一带一路"绿色发展相关研究、系列专题研讨、能力建设活动和绿色示范项目等。

强化对外投资企业的环境社会责任。 倡导对外投资企业要履行高标准的环境社会责任是包括联合国环境规划署、经济合作与发展组织、世界银行等在内的国际机构长期以来重点推动的工作。2000 年 7 月，联合国发布全球契约，旨在推动开展负责任的商业活动，推动全球企业实施符合环境、劳工等领域十大公认原则的战略，其中环境原则包括企业应对环境挑战未雨绸缪，主动增加对环保所承担的责任，鼓励无害环境技术的发展与推广。经济合作与发展组织于 1976 年制定了《跨国公司行为准则》，后经多次修订，该准则对可持续发展的重视程度逐渐提升，要求企业重视营运活动可能对环境造成的影响，强化环境管理系统。世界银行将环境影响作为筛选贷款项目的重要依据，要求申请项目必须提交符合标准的环评报告。

2013 年，中国商务部、环境保护部共同发布《对外投资合作环境保护指南》，倡导企业树立环保理念，依法履行环保责任，要求企业遵守东道国环保法规，履行环境影响评价、达标排放、环保应急管理等环保法律义务。2016 年 12 月，在环境保护部（现生态环境部）、外交部、国家发展改革委、商务部等部委的支持下，19 家来自能源、交通、制造、环保等行业的重点企业共同发起《履行企业环境责任，共建绿色"一带一路"》倡议，倡导在对外投资贸易活动中遵守环保法规、加强环境管理，助力绿色丝绸之路建设。2019 年 4 月，中国与英国、法国、新加坡、巴基斯坦、阿拉伯联合酋长国等有关国家和地区主要金融机构共同签署《"一带一路"绿色投资原则》，这一原则的签署标志着"一带一路"投资绿色化走向新的阶段。

区域环境治理能力不断提升。 2015 年 12 月，中非合作论坛约翰内斯堡峰会提出实施中非绿色发展合作计划，支持非洲实施 100 个清洁能源和野生动植物保护项目、环境友好型农业项目和智慧型城市建设项目，共同推进中非环境合作中心建设，增强非洲绿色、低碳、可持续发展能力；2018 年 1 月，中国与澜湄国家（包含缅甸、老挝、泰国、柬埔寨、越南）共同发布《澜沧江—湄公河合作五年行动计划（2018—2022）》，提出推进澜沧江—湄公河环境合作中心建设，制订并实施"绿色澜湄计划"，重点推动大气、水污染防治和生态系统管理合作；2018 年 7 月，中国生态环境部与柬埔寨环境部在金边签署了《共同设立中国—柬埔寨环境合作中心筹备办公室谅解备忘录》，并启动中国—柬埔寨环境合作中心筹备办公室，推动开展生活污水处理、生物多样性保护等相关领域的项目示范工作。为促进共建"一带一路"国家环保能力建设和人员交流，中国政府启动了绿色丝路使者计划和应对气候变化南南合作培训，每年支持 500 多名共建国家代表来华交流。主题涉及环境影响评价、大气污染防治、水污染防治等方面，参加活动的人员包括共建国家环境部门官员、青年学生、非政府组织志愿者、专家学者等。

积极推动将绿色经济理念融入"一带一路"倡议中。当前，发展绿色经济已经成为全球共识。联合国环境规划署已经在非洲、亚太地区、加勒比地区和拉丁美洲相关国家开展了绿色经济相关工作，包括与绿色经济有关的政策、战略、指标体系研究等，并且正在积极推动绿色经济理念融入绿色丝绸之路建设。中国工商银行发行首只"一带一路"银行间常态化合作机制（BRBR）绿色债券，并与欧洲复兴开发银行、法国东方汇理银行、日本瑞穗银行等 BRBR 机制相关成员共同发布"一带一路"绿色金融指数，深入推动"一带一路"绿色金融合作。中国光大集团与有关国家金融机构联合发起设立"一带一路"绿色投资基金，解决共建"一带一路"国家绿色股权投资不足、合作机制缺失等问题，推动共建国家绿色金融务实创新。

企业与环保非政府组织推动清洁高效技术在共建国家推广应用。中国大力推动清洁高效技术在"一带一路"项目中的应用。2016 年 9 月，科技部、国家发展改革委、外交部、商务部联合发布了《推进"一带一路"建设科技创新合作专项规划》，提出将节能减排理念充分融入各个重点科技合作领域，包括开展高效节水与节能农业等技术和农机装备的联合开发与示范，推广环境友好型和气候智慧型农业发展模式，推动适合沿线国家实际情况的太阳能、生物质能、风能、海洋能、水能等可再生能源合作以及煤、油、气等传统能源清洁高效利用技术的研发和示范，积极推动新能源汽车合作开发，加强海洋灾害、极端天气气候、地质灾害、洪旱灾害等数据共享、技术和经验推广等。中国商务部与联合国开发计划署签署在埃塞俄比亚、斯里兰卡的可再生能源三方合作项目协议。中国生态环境部与深圳市共建"一带一路"环境技术交流与转移中心（深圳），并在江苏宜兴、广西梧州等地设立环保产业国际合作示范基地，打造与共建国家开展环保产业技术合作的平台。

海洋环境合作稳步推进。中国已经与泰国、马来西亚、柬埔寨、印度、巴基斯坦等国建立了海洋合作机制。目前中泰气候与海洋生态系统联合实验室、中巴联合海洋科学研究中心、中马联合海洋研究中心建设都在积极推进，重点将在海洋与气候变化观测研究、海洋和海岸带环境保护、海洋资源开发利用、典型海洋生态系统保护与恢复、海洋濒危动物保护等多领域开展合作。

绿色金融为绿色丝绸之路建设提供保障。为支持"一带一路"建设，中国于2014年年底出资400亿美元成立丝路基金。2017年5月，中国政府宣布向丝路基金增资1 000亿元人民币。截至2019年11月，丝路基金已签约34个项目，承诺投资金额约123亿美元。丝路基金倡导绿色环保、可持续发展理念，积极支持绿色金融和绿色投资。

1.3 建设绿色"一带一路"，将绿色发展由机遇转变为现实

推动"一带一路"绿色发展需在"一带一路"建设中建立环境与发展的综合决策机制。将绿色发展与生态环境保护融入"一带一路"建设的各个方面，加强建立与2030年可持续发展目标接轨、符合国际合作模式的绿色化引导政策和可操作性强的实施指南，从而为共建"一带一路"国家实现可持续发展目标中的环境与社会目标做出直接贡献。

促进"一带一路"绿色发展需充分理解在建和规划中的"一带一路"项目对当地和全球环境以及可持续发展产生的影响。开展"一带一路"项目对当地和全球环境及可持续发展影响的专题分析，将帮助判断应当在未来几年中为哪些类型的项目提供更大的政策和资金支持。开展企业和投资者参与"一带一路"项目的正面和负面影响专题分析，能够推动投资行为转变，确保投资者做出环境友好和可持续的决策[1]。

"一带一路"的绿色化需推动落实绿色基础设施原则。联合国2030年可持续发展议程将建设可持续的基础设施确定为17项主要发展目标之一。投资与建设绿色、可持续的基础设施是共建"一带一路"国家应对气候变化、资源短缺、人口增长等全球性挑战的需要。"一带一路"要建设能促进低碳和环境可持续发展

---

[1] 中外对话网站. 牛津大学发起"一带一路"环境数据平台[EB/OL]. [2017-12-15]. https://chinadialogue.net/zh/1/43848/.

的基础设施，比如可再生能源发电厂以及公共交通系统，并在项目规划、建设和运营阶段充分考虑环境、社会与治理（ESG）要素。最大限度地发挥基础设施在实现可持续发展目标、提高经济效益的同时，兼顾环境和社会考量、增强自然灾害风险抵御能力等方面的积极作用。

在"一带一路"项目和相关国家推广绿色发展需对绿色化水平更高的项目给予融资支持。绿色金融将为改善环境、减缓与适应气候变化、促进资源保护和高效利用等提供支持，即为环境保护、清洁能源、绿色交通运输、绿色建筑和可持续农业等领域提供金融服务，包括项目投资与融资、项目运营和风险管理等。持续推动绿色金融实践能够引导资金流入绿色化水平更高的产业和项目。

"一带一路"绿色化需建设绿色价值链和绿色供应链体系。推进绿色生产、绿色采购和绿色消费，带动产业链上下游采取节能环保措施，以市场手段降低生态环境影响。

提高"一带一路"绿色化水平需推动节能环保产业和技术合作，提供有效的环境治理方案。通过环保技术最佳实践的交流与合作，加强共建"一带一路"国家污染防治技术能力建设，促进环保技术的转移和开发，帮助共建国家发展适合本国需求和实际情况的清洁产业，通过污染控制技术推动有效环境治理。

1.4 绿色"一带一路"对落实 2030 年可持续发展目标的潜在贡献

2015 年联合国通过的 2030 年可持续发展议程及《巴黎协定》带领国际社会迈入了可持续发展的新阶段。中国已做出建设绿色丝绸之路的承诺。建设绿色丝绸之路，顺应了国际社会绿色发展的潮流与趋势，与 2030 年可持续发展议程及《巴黎协定》的理念高度契合。绿色"一带一路"将绿色发展和生态环保理念融入"一带一路"建设的各方面和全过程，为共建国家实现 2030 年可持续发展议程设定的环境目标做出直接贡献，也为共建国家带来可持续发展的重要机遇。

推动生态环保政策沟通，加强可持续发展伙伴关系。政策沟通是绿色"一带一路"建设的基础，将推进绿色"一带一路"建设纳入联合国可持续发展目标框架，可以兼顾各个国家的不同利益需求和关切点，在政策层面与可持续发展目标实现对接，加强区域可持续发展政策的一致性，有效推动在共建"一带一路"国家间加强全球可持续发展伙伴关系（目标 17），并根据共建国家的不同需求将有关可持续发展项目融入各国政府的国家与地方可持续发展进程中。

在联合国可持续发展目标框架下推进绿色丝绸之路建设，还将有助于共建"一带一路"国家实现绿色、包容、可持续经济发展，避免发展中国家重走"先污染，后治理"的发展道路，处理好经济发展与生态环境保护的关系。作为绿色发展的倡导者，中国有能力促进共建"一带一路"国家的环境认同并为推动区域落实2030年可持续发展目标中的环境目标贡献中国方案。中国的生态文明理念可以提升绿色转型治理能力，绿色丝绸之路建设能够推动与共建国家共同开展全球生态文明建设，共建绿色命运共同体，帮助沿线发展中国家将可持续发展目标融入国家、地区发展规划和项目开发中，在项目全生命周期综合考虑环境、社会和经济等多方面因素。

中国将加强生态环保合作机制与平台建设，与共建国家开展政府间高层对话，利用中国—东盟、上海合作组织、澜沧江—湄公河、欧亚经济论坛、中非合作论坛、中阿合作论坛等合作机制，强化区域生态环保交流。政策沟通活动将有效推动共建"一带一路"国家间加强全球可持续发展伙伴关系（目标17.16），同时还有利于加强区域可持续发展政策的一致性（目标17.14），为各国创造平台协调可持续发展政策，开展经验交流，分享最佳案例。

防范设施联通环境风险，保护区域生态系统。绿色"一带一路"要求在基础设施建设过程中加强对建设中基础设施的绿色化工程，不断推进生态环保公共产品和环保基础设施建设，推进环保产业技术转移交流合作示范基地、环保产业园区建设，有利于基础设施升级，并保证在基础设施建设过程中更多地采用清洁和环保技术。"设施联通"中的绿色要素与多个可持续发展目标相关，该措施一方面可保护、恢复和促进可持续利用陆地和内陆的淡水生态系统及其服务（目标15.1）、减少自然栖息地的退化，遏制生物多样性丧失（目标15.5）；另一方面还可帮助有关国家升级基础设施，提高资源使用效率，更多的采用清洁和环保技术及产业流程（目标9.4）。

共建"一带一路"国家可持续基础设施的建设还将有助于打造一体化的生态环境风险防范和治理体系，共建可持续"一带一路"基础设施投资激励框架，推动建立"一带一路"可持续基础设施项目开放数据库，在环境和基础设施规划过程中落实最佳实践。

促进绿色贸易畅通，改善生产消费效率。在贸易领域，绿色"一带一路"将促进环境产品和服务贸易便利化，分享环境产品和服务合作的成功案例，提高环境服务市场开放水平，扩大大气污染治理、水污染防治等环境产品和服务进出口，其中绿色供应链国际合作是重要措施，通过建设"一带一路"绿色供应链合作平台，从生产、流通、消费的全产业链角度推动绿色发展。这些活动将通过贸易的手段提高共建国家的可持续生产和消费水平，有利于"一带一路"区域逐步改善消费和生产的资源使用效率（目标8.4）。

推动绿色资金融通，促进清洁技术投资。绿色金融能够为"一带一路"倡议保驾护航。绿色"一带一路"要通过金融工具全面认识和积极预防项目中的社会与环境风险，做好环境信息披露，加强项目环境风险管理，提高对外投资的绿色水平，"一带一路"沿线的绿色投资已经受到广泛关注，发展绿色产业前景广阔[①]。绿色金融可有效促进对能源基础设施和清洁能源技术的投资（目标7.A）[②]，支持从多渠道筹集额外财政资源用于沿线发展中国家（目标17.3）。例如，丝路基金从业务开展初期就贯彻绿色发展、绿色金融的理念，将注重绿色环保和可持续发展作为投资原则之一，重点推动清洁、可再生能源领域多层次互动，形成能源资源广泛合作。

以清洁能源为例，中国已成为全球最大的可再生能源市场，也是全球最大的清洁能源设备制造国。来自中国的清洁能源产业和技术正在共建"一带一路"国家中蓬勃发展，成为推动全球能源转型的主要驱动力之一（图1-1）。美国布鲁金斯学会指出，中国将彻底改变全球清洁能源产业，将对全球能源消费主要由化石燃料转向非化石燃料发挥关键性作用。尽管仍然需要大量煤炭和石油，但中国将在未来20年内向低碳发电领域和其他清洁能源技术投入6万多亿美元。中国已为基础设施投资发行近250亿美元的绿色债券，相关投资涵盖清洁能源、清洁交通、污染防控、节能技术等众多领域。

---

① 中国金融信息网． "一带一路"：推动绿色金融发展新契机[R/OL]．[2018-07-09]. http://greenfinance.xinhua08.com/a/20180709/1768414.shtml.
② 目标7.A：到2030年，加强国际合作，促进获取清洁能源的研究和技术，包括可再生能源、能效，以及先进和更清洁的化石燃料技术，并促进对能源基础设施和清洁能源技术的投资。

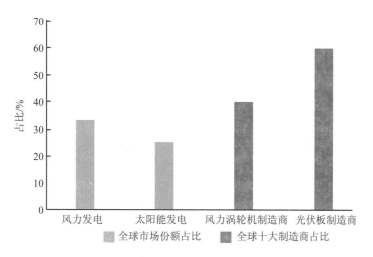

图 1-1 中国占全球清洁能源市场比例

数据来源：国际能源署。

加强环保民心相通，推动发展中国家能力建设。"民心相通"是"一带一路"建设的"关键基础"，只有实现了"民心相通"，绿色"一带一路"才可能真正在共建国家落地。绿色"一带一路"将加大绿色示范项目的支持力度，推动在环保政策、法律制度、人才交流、示范项目等方面开展合作交流，继续实施绿色丝路使者计划，加强共建国家环境管理人员和专业技术人才的互动与交流，推动环保技术和产业合作，提升共建国家的环保能力。上述活动和项目将有效支持发展中国家加强科学和技术能力，采用更可持续的生产和消费模式（目标12.A），促进发展中国家开发以及向其转让、传播和推广环境友好型技术（目标17.7），通过南南合作加强国际社会对在发展中国家开展高效和有针对性的能力建设活动的支持力度（目标17.9）。

2 绿色"一带一路"建设中的机遇与挑战

2.1 机遇

作为绿色发展的倡导者,中国有能力促进共建"一带一路"国家的环境认同,并为推动区域落实 2030 年可持续发展目标中的环境目标贡献中国方案。将中国生态文明建设的重要理念和实践成果融入"一带一路"建设中,不但丰富了"一带一路"建设的内涵,而且是"一带一路"倡议走向高质量发展阶段的必然选择。建设绿色丝绸之路能够为共建国家提供绿色转型的样板,以生态环境保护促进经济高质量发展。

绿色丝绸之路将为共建"一带一路"国家带来绿色发展和加强能力建设的巨大机遇,主要包括:

- 推动环保公众意识提升,减少资源消耗,促进绿色产业和低碳生活方式;
- 帮助共建"一带一路"国家将可持续发展目标融入国家、区域规划和具体项目开发过程中;
- 在"一带一路"项目中推广较严格的生态环境标准;
- 在促进可持续发展的同时为"一带一路"倡议打造一体化的生态环境风险防范和治理体系;
- 与共建"一带一路"国家的政策制定者合作,共建可持续"一带一路"基础设施投资激励框架,推动建立"一带一路"可持续基础设施项目开放数据库,在基础设施规划过程中落实最佳实践;
- 建立跨领域的绿色"一带一路"学习与领导力平台,提高官员及公众对环境风险、发展机遇及其应对方法的关注和认知;
- 为创造开放的、完善的政策环境以及持续推动透明度提升提供支持。

2.2 挑战

21世纪以来，可持续发展成为国际社会主题，国际环境与发展领域取得了很多重要进展，但与此同时，人类给地球带来的环境压力还在持续增加，支撑可持续发展目标实现的三大维度中，环境维度仍然落后于社会和经济发展进程，国际环境与发展领域也还有"开倒车"的现象。"一带一路"倡议包含有史以来规模最大的基础设施项目合作，在创造大量机遇的同时，也给投资者、可持续发展和自然资源带来了挑战（图2-1），与此同时，虽然可持续发展是国际社会的主题，但共建国家生态环保能力不足、国际合作项目的复杂性均给绿色丝绸之路建设带来了一系列挑战。

在绿色理念方面，"一带一路"沿线很多国家发展水平还不高，仍将经济迅速发展作为当务之急，绿色发展问题并不急迫。引入低碳科技和加强生态环保要求还未得到充分重视或仅处于初期阶段。

在政策支持与监控评估方面，"一带一路"项目大多涉及多个国家，在计划、设计、建设、运营和评估项目的过程中面对不同的标准和流程。投资可持续基础设施的商业环境并不明朗，在部分情况下，当地法律和技术标准十分模糊甚至完全缺失；可持续标准与评估方法的数量和种类不足，从而使金融投资者很难确保全部投资都流向可持续基础设施；对于一些可持续基础设施来说，由于可持续项目缺少收益流或公共政策激励，收益回报率往往不理想，影响了投资者的积极性。

在信息公开与透明度方面，跨国项目实施起来通常都十分困难。由于"一带一路"项目的规划、设计和落实大多是分散的，而非从整体出发，因此相关数据也难以集中和定位。"一带一路"项目具有高度复杂性，因此在执行过程中就必须有详细规划，并保持极高透明度。

在绿色金融与绿色投资方面，截至目前，绿色金融与绿色投资还没有引起金融和私营领域参与者和其他利益相关方的足够关注。绿色项目暂无一个普遍认可的定义，因此较难界定在现有的绿色投资中哪些可以给予融资。与对应的风险相比，"绿色"项目的收益率和投资回报率都十分有限。

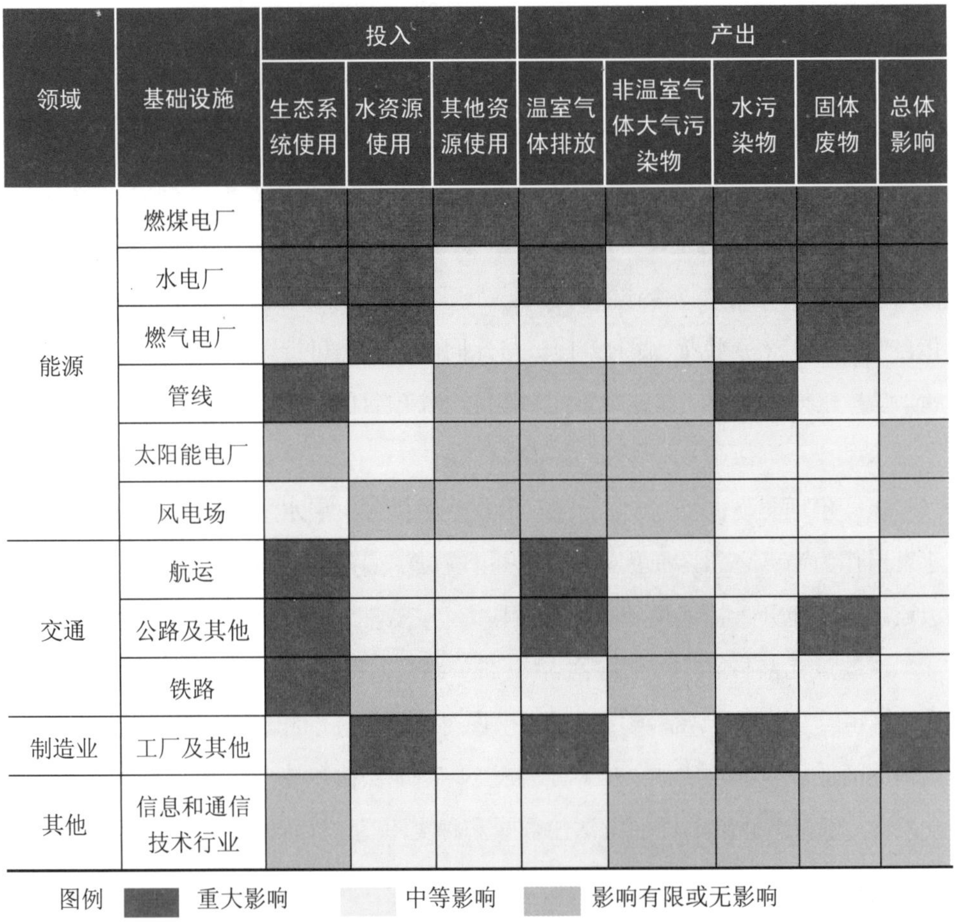

领域	基础设施	投入			产出				
		生态系统使用	水资源使用	其他资源使用	温室气体排放	非温室气体大气污染物	水污染物	固体废物	总体影响
能源	燃煤电厂	重大	重大	中等	重大	重大	重大	重大	重大
	水电厂	重大	重大	有限	有限	有限	重大	有限	重大
	燃气电厂	有限	中等	有限	重大	中等	中等	中等	中等
	管线	重大	有限	有限	有限	有限	中等	有限	中等
	太阳能电厂	中等	有限	中等	有限	有限	有限	有限	中等
	风电场	中等	有限	有限	有限	有限	有限	有限	有限
交通	航运	重大	有限	有限	重大	重大	重大	有限	重大
	公路及其他	中等	有限	有限	重大	中等	有限	重大	中等
	铁路	中等	有限	中等	中等	有限	有限	有限	中等
制造业	工厂及其他	有限	重大	中等	中等	中等	重大	重大	中等
其他	信息和通信技术行业	有限	有限	有限	有限	有限	有限	有限	有限

图例　■ 重大影响　□ 中等影响　▨ 影响有限或无影响

图 2-1　"一带一路"基础设施项目生态足迹

资料来源：汇丰银行，世界自然基金会."一带一路"倡议绿色化报告，2017。

3 "一带一路"建设重要关注问题研究

3.1 推动绿色"一带一路"建设形成长效机制

绿色"一带一路"建设的长效机制包括战略对接机制、规划嵌入机制、项目审评机制等方面。

3.1.1 战略对接机制研究

(1) 推进绿色"一带一路"建设战略对接机制的建设现状

战略对接机制是绿色"一带一路"建设的重要基础。过去6年,围绕绿色"一带一路"、可持续发展等主题,在中国和许多共建"一带一路"国家举办了一系列多领域、多层次的双(多)边对话交流活动,吸引了中国和共建国家政府官员、智库专家、媒体代表和企业家以及有关国际组织代表的积极参与,包括欧亚经济论坛生态与环保合作分会、中国—阿拉伯国家环境合作论坛、中国—东盟环境合作论坛等。

2015年以来,我国与多个共建"一带一路"国家签署了近50份双边和多边生态环境合作文件,根据双方合作意愿,在环境政策对话与信息共享、污染防治与环境管理、生态系统管理与生物多样性保护、环境保护产业与技术、环境监测、环境教育和公众参与以及经双方同意的与保护和改善环境有关的其他领域进行合作,并通过分享信息与资料,互派专家学者和代表团,组织有关研讨会、培训班及联合会议等方式进行合作交流。

(2) 推进绿色"一带一路"建设战略对接机制存在的问题

一是各国对生态环保问题认识的差异增加了对接难度。共建"一带一路"国家由于发展阶段不同,对生态建设和环境保护的认识和资源投入力度各不相同,环境管理法规和政策也不一致,有些国家的相关法律法规和技术标准不够完善,从而导致各国开展绿色"一带一路"倡议对接时不能采取统一的策略和标准,加大了对接难度。

二是目前没有在绿色"一带一路"建设框架下设立专门推动战略对接的机制。现有战略对接主要依托"一带一路"国际合作高峰论坛、中国—东盟、上海合作组织和金砖国家等合作机制下的环境部长会议开展,生态环境相关部门也积极推动与共建国家签署环境合作文件,但尚未形成绿色"一带一路"建设框架下的专门机制。

3.1.2 规划嵌入机制研究

(1)推进绿色"一带一路"建设相关规划编制情况

作为《中国国民经济和社会发展第十三个五年规划》的重要专项规划,《"十三五"生态环境保护规划》对推进"一带一路"绿色化建设提出了明确要求,并专门设置了"推进'一带一路'绿色化建设"相关内容,对 2016—2020 年绿色"一带一路"建设各项工作做了统筹安排。

此外,在"一带一路"建设六大经济走廊过程中,2016 年 6 月出台的《建设中蒙俄经济走廊规划纲要》中也加入了加强生态环保领域合作的任务,包括研究建立信息共享平台的可能性、扩大防灾减灾合作、开展生态环境保护领域的技术交流合作等。2015 年 11 月出台的《中国—中东欧国家合作中期规划》中也涉及节能环保产业合作、可持续采矿等环保和可持续发展合作的内容。

(2)绿色"一带一路"规划嵌入机制存在的不足

一是生态环境保护合作还没有成为"一带一路"建设有关合作规划的重点内容。在现有的"一带一路"建设合作规划中,无论从篇幅还是重要性表述来看,生态环境保护和绿色发展还没有成为重点内容,如前述《中国—中东欧国家合作中期规划》的 58 条合作内容中,只有第 39 条和第 40 条两处与生态环境保护合作相关。而基础设施互联互通、国际产能合作、境外经贸合作区建设是现阶段合作重点。

二是尚未制定系统全面的"一带一路"建设框架下的双(多)边生态环境保护合作规划。目前,我国与共建国家主要以签署谅解备忘录或政府间协议的方式明确开展"一带一路"建设合作,与部分国家(如哈萨克斯坦)编制了部分领域(如国际产能合作)规划,还没有编制生态环境保护和绿色发展合作的专项规划。

3.1.3 项目审评机制研究

（1）我国对"一带一路"建设项目管理现状

目前"一带一路"建设相关合作项目按照原有管理方式和渠道，由相关部门或行业组织负责推进、企业负责具体实施。国家发展改革委等有关部门支持以企业为主体、以市场为导向、按商业原则和国际惯例的对外投资项目，尤其支持企业投资和经营"一带一路"建设及国际产能合作项目。根据 2015 年国务院发布的《国务院关于推进国际产能和装备制造合作的指导意见》（国发〔2015〕30 号）精神，铁路（扩展到基础设施）、电力、汽车、通信等是国际产能合作重点行业。

（2）"一带一路"建设项目生态环境风险评估存在的问题

生态环境影响和风险在项目评估决策中的重要性进一步提高。目前，对"一带一路"建设项目的评估已考虑经济可行性、技术合理性以及对当地经济社会发展的带动作用等因素，生态环境影响等也作为风险因素纳入考虑评估。但由于各行业部门和金融机构对项目的评估方法和标准没有统一规范，生态环境影响和风险在项目评估决策中的权重还有提升空间。"一带一路"绿色发展国际联盟于 2019 年年底启动了"一带一路"项目绿色发展指南研究项目，旨在制定"一带一路"项目分级分类指南，评估项目生态环境和气候影响，为利益相关方进一步识别和应对海外投资的生态环境风险提供指引。

3.2 "一带一路"建设中绿色金融的战略布局与实施机制

绿色金融助力绿色"一带一路"建设主要体现在两个方面：一是环境风险管理问题，即金融机构如何加强对环境保护、生态保育和应对气候变化等因素在投资决策中的考量；二是绿色投融资问题，即金融机构如何加大对绿色、低碳和循环经济项目的投资力度，并提供相应的程序上的便利。这是金融机构践行绿色金融或绿色信贷的基本内容，在 2012 年 2 月中国银监会（2018 年与中国保险监督管理委员会合并组建为中国银行保险监督管理委员会，简称中国银保监会）发布的《绿色信贷指引》和 2016 年 8 月由中国人民银行、国家发展改革委等七部门发布的《关于构建绿色金融体系的指导意见》中均有体现。中国金融机构在支持"一带一路"倡议时，不仅仅是简单地执行上述两方面内容，更重要的是有别于在中国境内实施的绿色金融,注重如何通过金融在"一带一路"建设中体现联合国2030年可持续发展目标,同时将中国的绿色金融实践与共建"一带一路"国家和地区的特定国情和投资环境相结合。

3.2.1 国际上绿色金融的良好实践及借鉴

国际上对金融机构的环境风险管理要求主要来自以下四个方面:

一是国际准则与国际合作。许多国际机构和地区性的多边机构都制定或推动形成了国际性的环境与社会治理标准,并大力推进各国和机构之间的对话,如《联合国全球契约》《赤道原则》《联合国负责任投资原则》等。

二是多边发展性金融机构制订的标准。发展性金融机构,基于其经营宗旨或遵守的国际标准,都推出了自己的标准与指导方针,例如泛美开发银行、世界银行集团、亚洲开发银行和非洲开发银行等。这些标准与实施工具在很大程度上被各国政府和金融机构学习和借鉴。

三是国家政策与法规。这一机制可以有不同的形式:环保法规将划定界限,通常会规定项目的环境评估,划定保护区域,确定污染限值等。此外,近年来针对金融机构的绿色监管已日趋严格和完善,包括对个别行业投资活动的限制,以及对绿色投资的激励措施等,如2012年中国银监会发布的《绿色信贷指引》及相关政策、2017年巴西中央银行制定的《社会和环境责任政策》、2012年尼日利亚中央银行制定的《可持续银行业原则》、2017年孟加拉国中央银行制定的《环境社会风险管理指引》等。

四是商业性金融机构的自发承诺。商业性金融机构由于自身可持续发展的需求或由上述国内外政策的推动,承诺遵守环境与社会标准,更好地管理自身环境和社会绩效,并在投资经营活动中加强对环境风险的管理。

不管形式来源如何,环境风险管理最实质性的表现是金融机构将可持续发展作为机构战略重点,环境因素有机嵌入投资和授信决策流程,并公开披露环境敏感行业的投资和信贷政策。

而对于绿色投资的引导,也有以下四种形式:

一是由国家制订的绿色投资目录或标准,这是中国鼓励绿色投资的主要方式。

二是由国家出资成立专门的绿色投资机构来引导市场和商业性机构进行绿色投资，比较典型的是英国绿色投资银行的模式。

三是由金融机构根据自身战略和宗旨制订的投资优先考虑和重点发展领域。如世界银行在其 2018 年年报中强调世界银行在三个重要领域向客户国提供支持：促进可持续、包容性的经济增长；更多且更有效地推动以人为本的投资；建立应对全球经济脆弱性、冲击以及危机的韧性。而汇丰银行、花旗银行等商业金融机构也制定了加大可再生能源、低碳基础设施建设投资力度的战略。

四是由行业协会、研究咨询机构等发起并获得广泛认可的标准。这种类型比较典型的是绿色债券，由国际资本市场协会（ICMA）制定的《绿色债券原则》和由气候债券倡议组织制定的一些标准和规范，已逐步被金融机构接受和遵守。

同样，积极开展绿色投资的实质性表现是在风险可控的前提下，通过产品创新、机制创新和程序创新，实现对绿色、低碳和循环经济项目的投资额增长。

3.2.2　环境风险管理在绿色"一带一路"建设中的制度安排

中国政府一贯重视中资企业和金融机构在境外经营中的环境和社会风险[①]。

中国银监会于 2012 年 2 月发布的《绿色信贷指引》，推动银行业金融机构以绿色信贷为抓手，积极调整信贷结构，有效防范环境与社会风险，更好地服务实体经济，促进经济发展方式转变和经济结构调整；于 2014 年 6 月发布的《绿色信贷实施情况关键评价指标》，落实了《绿色信贷指引》等监管规定，从组织管理、政策制度及能力建设、流程管理、内控管理与信息披露、监督检查等多方面制定了定量和定性指标，指导银行业金融机构每年开展绿色信贷实施情况自评价工作；于 2017 年 1 月发布了《关于规范银行业服务企业走出去　加强风险防控的指导意见》，其中第五条要求银行业金融机构应持续提升全面风险管理能力，包括环境和社会风险管理，尤其要求银行业金融机构对能源资源、农林牧渔、重大基础设施及工程承包领域的环境和社会风险，在提供项目融资

---

① 新华网. 中国对外投资环境风险管理倡议发布[R/OL]. [2017-09-06]. http://www.xinhuanet.com/money/2017-09/06/c_1121612075.htm.

及贸易融资时给予特别关注。

2016年8月，中国人民银行、财政部、环境保护部（现生态环境部）等七部门联合印发了《关于构建绿色金融体系的指导意见》，从环境信息披露机制与市场监管规则两个方面强化我国绿色金融政策基础。该意见第八部分提出广泛开展绿色金融领域的国际合作、推动提升对外投资绿色水平，鼓励和支持我国金融机构、非金融企业和我国参与的多边开发性机构在"一带一路"的项目投资和建设中强化绿色发展理念，在环境风险管理、环境信息披露、绿色供应链管理和环境污染责任保险等方面提升绿色低碳嵌入水平。

2017年5月，环境保护部发布的《"一带一路"生态环境保护合作规划》将推进绿色投融资作为建设绿色"一带一路"的重要组成部分。"一带一路"框架下绿色金融体系的构建应从国家战略部署出发，坚持可持续发展理念，鼓励和支持更多金融资源投向绿色产业，借助绿色债券、绿色信贷等金融工具控制并减少污染性投资，健全环境、社会和治理方面的标准、指引和流程管理，督促更多金融机构有效识别、评估、控制投融资活动中的环境和社会风险，形成促进经济向绿色化转型的制度合力。

上述文件都对金融机构在境外经营和投资活动中的"绿色化"和环境风险防范形成严格且详尽的规定和要求。但从实施效果来看，目前还存在一些问题，主要表现在：

一是绿色环保标准尚未完全统一，投资风险增加。目前，中国金融机构普遍未参照国际通用准则，在国际融资项目中所执行的环境政策主要是国内绿色信贷政策的延伸，一般是中国的法律法规或者参照项目所在国的标准，可能出现适用局限性。不同国家的政治、社会、文化、法律制度及价值观不同，投资环境迥异，国际投资合作面临着较大的不确定性。

二是绿色金融政策框架需完善，影响实施效果。作为"一带一路"建设中绿色金融的主要引导者，政府部门应对绿色金融境外适用的细化措施进行完善。各金融机构的执行标准不尽相同，影响了绿色投融资业务的具体开展和统计数据的质量。同时，环境数据可用性不足、环境风险分析方法缺乏、信息披露平台不健全等，都是制约绿色金融发展的技术性因素。在项目审批环节，有关部门对境外投资项目的环境因素考虑不足，已发布的相关政策实效性和约束力不强，难以有效引导对外直接投资中环境保护措施的实施。

三是绿色金融政策执行需细化，增强可操作性。作为"一带一路"建设的重要参与者，中国金融机构境外项目环保和社会管理在整体上与国际标准仍有一定的差距，与国际准则或惯例的匹配对接程度还可以进一步提高。在政策执行过程中，金融机构多数以合规评价来代替环境风险定量评价，通过"一票否决制"和"名单管理"等方式将环境政策纳入融资决策，未制定具体有效、明确量化的信贷实施细则，也没有针对境外投资环境绩效考核的明确标准。

四是部分企业环保意识仍有待提高，可继续强化全程监管。在实施绿色金融过程中，企业环保意识的提升与政府的引导、银行的参与相辅相成。目前，大多数参与"一带一路"的中国企业在环境保护及社会责任方面已积累了相对丰富的经验，但仍有部分企业环保意识亟待加强。中国已经在对外绿色投融资方面出台了一系列政策、指导和建议，但对非绿色投融资行为的约束力不足，尚未形成对外投资管理环境监管体系，涉外环境法律法规缺位，尤其是缺乏规制企业对外投资行为的专项环境法律法规。加强"一带一路"投资过程中的环境风险管理，重点是加强法律和行政监督，对因环境问题引发商业损失进行追责，加大处罚力度。

五是引导性资金支持力度不够，对市场的吸引力不足。绿色投融资涉及的环境风险评估、碳交易等十分复杂且不断更新的专业技术，对金融机构的风险评估和管理工作提出了非常高的要求。因此，应当先配备一定的引导性资金及能力建设资金，支持具备经验的相关政府部门从事专项资金管理和能力建设等相关工作，并逐步吸引市场机构加入。

3.2.3 绿色投资在"一带一路"中的制度安排

目前中国针对国内绿色投资的标准制订非常详细，其中包括中国银监会的《绿色信贷统计制度》、中国人民银行的《绿色债券支持项目目录》、国家发展改革委的《绿色债券发行指引》等。其中《绿色信贷统计制度》中，除11大类绿色产业和项目目录外，明确"采用国际惯例或国际标准的境外项目"也在绿色信贷统计范围内。由中国工商银行、中国农业银行、中国银行、兴业银行等在境外发行的绿色债券所得款项用途也符合国内绿色债券的发行要求。

目前"一带一路"的绿色项目投资还刚刚起步，成功落地项目有限，主要存在四个方面的问题：

一是"一带一路"绿色项目的标准。"一带一路"绿色项目标准有三个问题需明确。第一，绿色项目的范围，是包括所有的绿色、低碳、循环经济的内容，还是侧重选择一些重要行业和重要地区；第二，绿色标准的选择，是以国内绿色金融、绿色信贷为标准，还是以国际机构、多边银行及多数商业金融机构的良好实践为标准，或者是以投资所在国标准为基础，或者是重新制定绿色"一带一路"项目标准；第三，绿色标准的量化评价问题，绿色项目标准是以目前通行的项目类型和性质认定，还是以项目的效果影响和绿色贡献度认定。

二是共建"一带一路"国家的投资环境和风险防范。"一带一路"项目投资难落地的"瓶颈"问题是共建国家的投资环境不佳，政治风险、市场风险、汇率风险等难以把控。究其原因：第一，共建"一带一路"国家的政治、社会、宗教、文化等情况差别很大；第二，金融机构缺乏对非金融风险管理的经验积累和人才储备；第三，我国与相关国家之间的经贸协调机制不够完善。

三是"一带一路"绿色项目的商业回报。项目的成本高、投入大、短期收益率低、项目周期长也是绿色项目难以获得资金支持的重要原因。在"一带一路"建设中，中国金融机构对境外项目缺乏有效信息和正确认知，给绿色项目回报又增加了不确定性。

四是"一带一路"绿色项目投资的激励机制。"一带一路"绿色项目投资不足的根本原因在于激励不足。金融机构投资绿色"一带一路"项目难以得到项目所在国的区别性优惠，也难以得到额外的商业利益。因此，需对绿色"一带一路"项目通过风险补偿、信用担保、税收优惠、信贷贴息等综合配套政策进行激励和扶持，充分调动金融机构积极性，增加绿色信贷的执行动力。

3.3 "一带一路"与构建更具包容性的全球绿色价值链

3.3.1 打造绿色价值链在构建绿色"一带一路"中的重要意义

绿色价值链的核心内涵。近40年来，随着贸易投资的自由化、便利化水平不断提高，国际分工由传统的产业间分工转变为产业内分工。随着信息通信技术革新带来的信息沟通与跨境协调成本锐减，产品跨国制造现象越来越普遍，每个分工环节带来的价值增值遍布于世界各个角落,推动了全球价值链(Global Value Chains,GVCs)的形成。发达国家和发展中国家以不同的分工方式参与到价值生产、分配和再分配的过程中,包含了从设计、研发、生产、运输、消费再到回收利用在内的整个价值增值过程[1]。绿色全球价值链(Green Global Value Chains,GGVC)将绿色发展理念融入全球价值链,注重每一环节价值增值所产生的环境和气候影响,形成包括绿色设计、绿色采购、绿色生产、绿色产品、绿色销售、绿色消费、绿色回收及绿色材料等在内的闭合链条,并在全球生产重新布局的大背景下,研究伴随各个价值增值环节跨国转移背后的环境足迹。

"一带一路"建设为构建更具包容性的绿色全球价值链提供了机遇。"一带一路"建设,一方面,通过互联互通将不同发展阶段、不同历史文化背景的国家紧密连接在了一起,构建了国际化的商品、服务融通平台。另一方面,共建"一带一路"国家处于不同的经济发展阶段,发展阶段异质性导致面对的环境与发展问题不尽相同,单一的环境规则无法适用共建"一带一路"国家的集体诉求。这就需要以互通互信为基础,依托"一带一路"倡议,充分考虑不同国家在不同发展阶段的实际需求,结合不同国家的生态环境现状,通过"规则治理"确立多领域、多层次的环境标准。

构建更具包容性的绿色全球价值链对于"一带一路"建设的深入推进至关重要。"一带一路"建设所倡导的产能合作是以互利共赢为理念和原则,谋求共建"一带一路"国家和地区的健康发展与共同繁荣。绿色全球价值链作为绿色"一带一路"建设的重要着力点,倡导升级版的绿色全球分工,在协力降低环境负担的同时,各环节公平合理占有产业链收益,最终形成经济、社会、环境多维度共赢的局面,从而保障"一带一路"建设的深入推进和可持续发展[2]。

3.3.2 打造绿色价值链的内在必然性

环境作为全球性公共物品,是全球治理的重要内容。在全球生产碎片化的发展背景下,构建更具包容性的全球绿色价值链,是打造绿色"一带一路"的关键。打造全球绿色价值链,有助于降低发展中国家产业发展的资源消耗强度,减少对

---

[1] 周亚敏. 构建更具包容性的全球绿色价值链[N]. 人民日报,2017-05-17.
[2] 中国环境与发展国际合作委员会. 中国在全球绿色价值链中的作用[R]. 2016.

资源型大宗商品的外部依赖，助力区域可持续发展。未来通过加强南南合作、实施"一带一路"倡议打造全球绿色价值链可能成为发展中国家实现自身发展目标和履行国际义务的重要手段。此外，构建协调、包容、绿色、可持续发展的全球价值链是非常紧迫的，需要各国在战略高度重新思考国际投资、生产、贸易与合作，用绿色和可持续发展的理念重塑全球价值链，这对"一带一路"沿线国家及企业能力的提高和政策协调提出了新的要求。

4 "一带一路"绿色发展案例研究

4.1 巴基斯坦、斯里兰卡实地调研

2019年2月,国合会绿色"一带一路"与2030年可持续发展议程专题政策研究项目组赴巴基斯坦和斯里兰卡进行实地调研。目的是进一步了解巴基斯坦、斯里兰卡两国绿色"一带一路"建设现状,识别"一带一路"项目的环境影响、所面临的挑战以及评估方法,清晰认识两国绿色"一带一路"合作的态度及需求,总结现有"一带一路"项目的优秀实践,寻求新的机遇,通过政策措施推动"一带一路"绿色发展。

调研发现,两国总体认可中国项目环保工作。无论是两国政府还是两国研究机构或社会组织基本上对"一带一路"项目的环保工作表示认可,认为能较好地遵照当地的环境法规。究其原因:一方面,两国政府对项目本就有明确的环境管理规定;另一方面,"一带一路"项目合作机制是双方共商共建,当地合作企业也有较高环境保护的诉求和考虑。与此同时,也有一些社会组织认为,项目在环境保护领域可以做得更好,如在巴基斯坦的欧盟机构就希望中巴经济走廊项目实施过程中的环境信息更加透明公开。

4.1.1 巴基斯坦"一带一路"项目生态环保情况

(1)基本情况

巴基斯坦北部地区拥有全球公认的各种脆弱生态系统,包括温带落叶林、针叶高山森林、苔原和草原等生态系统。这些地区的山区生态系统还包括冰川,冰川是当地和下游社区最重要的淡水来源,是区域生物多样性的基石。这些生态系统对于高海拔地区的社区生计至关重要。贫困和对自然资源的极度依赖,以及极端天气和气候变化相关的挑战,包括冰川融化和冰川湖突发洪水,使得当地社区的脆弱性加剧。

中巴经济走廊是"一带一路"倡议的重要组成部分。中巴经济走廊北起喀什，南至巴基斯坦瓜达尔港，全长 3 000 km，对中巴关系有着核心影响，是"一带一路"六大经济走廊中的旗舰项目，被认为是该区域的"游戏规则改变者"。中巴经济走廊是一个加强区域一体化和贸易流动的区域连通性框架，涉及发展交通、能源、工业和其他形式的基础设施。中巴经济走廊由于起源于巴基斯坦北部地区，又被称为"中巴经济走廊的门户"，具有巨大的潜力，特别是在水电和生态旅游方面。这些地区还有丰富的自然资源，包括森林、水、冰川、生物多样性和矿物资源。

中巴经济走廊投资的关键支柱主要包括四个方面，即瓜达尔港、能源项目、基础设施建设和产业合作。中巴经济走廊的基础设施建设投资达 620 亿美元，如果不能合理规划，将对当地，尤其是脆弱的高山地区的生物多样性、居民生计产生严重的负面影响。中巴经济走廊建设的经济收益是显著的，但还是需要关注各类受益者以及项目在地方层面产生的影响，尤其是对当地社区的影响。中巴经济走廊的启动将对社区生计和脆弱的生物多样性和生态系统带来一定挑战。

巴基斯坦环境主管部门为气候变化部环境保护局，最核心的环境管理法律法规《环境保护法》规定，对于可能造成环境损害的项目进行综合环境影响评价，排污量不能超过国家环境质量标准。中巴经济走廊有较为完善的协商机制，无论是规划合作方案还是实施具体项目，都由中巴双方平等协商、充分论证、共同实施，相关项目严格执行巴方的环境法规，编写环境影响评价报告书，在实施中确保符合或优于国家环境质量标准。在未来的合作项目规划中，将会更加关注绿色和民生，更多考虑水电、天然气等清洁能源项目。

（2）对中巴经济走廊绿色发展的建议

在中巴经济走廊的建设过程中，必须尽早将环境方面的考量纳入规划。例如，在降低运输成本的同时，中巴经济走廊所需的公路、铁路和管道建设可能会破坏自然生态系统，导致该地区的生物多样性丧失。其他建设项目，如建设水电站，也可能导致栖息地破碎化、森林退化、地下水污染和土壤污染。因此，在基础设施规划初期及建设全过程中，需要分析可能产生的生态环境影响，统筹管理环境风险，制定完整的环境管理措施。

2016 年 2 月，巴基斯坦议会将 SDGs 纳入国家发展目标，随后将 SDGs 纳入国家发展框架和《巴基斯坦愿景2025》，并明确了推动可持续发展的重点目标、基线和指数。"一带一路"倡议和中巴经济走廊的绿色发展与SDG9的落实有直接关系，如"促进包容可持续工业化，到2030年，根据各国国情，大幅提高工业在就业和国内生产总值中的比例，使最不发达国家的这一比例翻番"。此外，"一带一路"倡议和中巴经济走廊的绿色发展还将为其他SDGs具体目标的实现提供支持。

为务实推动中巴经济走廊的绿色发展以及绿色"一带一路"建设，中巴两国政府可以展开联合行动，对中巴经济走廊进行战略性环境评估，保证所有项目的环境影响评估都符合一定的标准，并确保相关环境管理措施的落实。鉴于建设区域内生物多样性热点地区较多，应将关注点放在重点生物多样性区域上，针对这些区域采取特殊的保护措施。例如，世界自然保护联盟（IUCN）正在与巴基斯坦政府合作，在巴北部地区开展项目，旨在研究中巴经济走廊对该区域的生物多样性的影响。此外，还可以探讨"互联网+"概念下如何促进公众积极参与到生态系统修复、退耕还林以及碳减排中。

4.1.2 斯里兰卡"一带一路"项目生态环保情况

斯里兰卡有"印度洋十字路口"之称，其南部汉班托塔港离印度洋主航道只有10 n mile，是连接中东、欧洲、非洲至东亚大陆的海运航线必经之地，也是"21世纪海上丝绸之路"的关键节点之一。

"一带一路"倡议实施以来，中斯两国开展了涉及港口、电信、供水、公路等多个领域的务实合作，项目总体上是基于当地经济社会发展的迫切需要和其历届政府在充分科学论证的基础上提出的。2017年，中斯两国政府签署了《中华人民共和国政府和斯里兰卡民主社会主义共和国政府关于促进投资与经济技术合作框架协议》，中国商务部与斯里兰卡发展战略和国际贸易部签署了《中国—斯里兰卡投资与经济技术合作发展中长期规划纲要》。在此背景下，一系列交通基础设施项目正在稳步实施，科伦坡港口城、汉班托塔港运营、汉班托塔临港工业园等重大合作项目正在积极推进。中斯合作有关项目为斯里兰卡创造了10余万个就业机会，培训了上万名技术与管理人才。

根据斯里兰卡的《国家环境法》和《环境影响评价法》，在斯里兰卡的项目要遵守严格的环境影响评价程序，项目实施方需编制环境影响评价报告，经中央环境署或海岸保护和海洋资源管理局等涉环境相关主管部门审批认可后，项目方可启动。中国有关企业参与的"一带一路"项目基本上能严格执行斯里兰卡相关环境保护政策和制度。

4.1.3 绿色发展案例对绿色"一带一路"建设的主要启示

总体而言，所有"一带一路"项目都应遵循以下原则。

（1）"一带一路"倡议与2030年可持续发展议程协同发展

- "一带一路"倡议的目标与2030年可持续发展议程和《巴黎协定》高度契合。"一带一路"项目的开发必须在不超过环境和资源承载能力的前提下进行。
- "一带一路"项目的发展应从第一阶段向第二阶段转型。第一阶段以大型基础设施项目为主，第二阶段则以建设经济特区、社会投资和解决环境问题为主。
- 所有新成立的"一带一路"相关组织和网络（包括"一带一路"绿色发展国际联盟、"一带一路"绿色制冷倡议、"一带一路"绿色照明倡议和"一带一路"绿色"走出去"倡议）都必须以加强现有多边机制和促进2030年可持续发展议程落实为目的。
- 中国投资者和项目实施企业应落实绿色金融与投资实践及企业社会责任标准。由27家国际大型金融机构支持设立的"一带一路"绿色投资原则就是一个良好的开端。
- 应加强各国落实2030年可持续发展议程和落实"一带一路"相关政策的协同发展，同时充分考虑各国实际情况。

（2）落实相关原则，确保项目从初始便注重绿色发展

- 需要制定相应的原则，确保新项目从初始阶段就以绿色发展为目标。包括帮助投资者树立环境保护理念；针对国内开发性银行制定指南，引导金融机构充分考虑生态环境风险；在5年内开始试点项目的建设，确保基础设施的设计符合生态原则（如采用基于自然的基础设施解决方案）。

- 需要采取全面、一体化的评估方法对项目总体环境影响进行评估,并因地制宜制定有效合理的环境管理措施。中国经过近些年的发展,形成了可推广的绿色发展经验和方法论,如长江经济带建设采取了坚持生态保护、高质量发展的理念,强化生态优先绿色发展的环境管理措施,实现了生态环境保护和经济发展之间的"共赢"。
- 应将经济特区建设成环境经济特区。考虑推出零排放经济特区标准。
- 应对项目进行事前和事后评估,组建包括学术机构和智库专家在内的国际团队,就项目对所在国有何社会和环境影响,以及是否与2030年可持续发展议程及其落实相一致进行评估。
- 应通过促进南南合作解决"一带一路"实施中的潜在生态环境问题。

研究针对三个不同层面的绿色发展提出建议。

在"一带一路"项目层面:

- "一带一路"项目应当避免、减少和补偿对当地环境和社会的负面影响(通过提供环境和社会保护措施),包括与利益相关方进行磋商。
- 用国际认可的方法对"一带一路"项目的环境影响进行评估,包括长期影响和潜在的不可逆影响。
- 对项目与2030年可持续发展议程和《巴黎协定》在相关国家的落实是否一致进行评估。避免化石能源投资和非弹性基础设施的路径依赖和锁定效应。
- 积极应对复杂和跨境项目所带来的挑战。

在经济走廊/区域项目层面:

- 确保环境可持续性、累积影响评估和政策连贯性。
- 了解经济走廊/区域项目工程如何融入相关国家的结构性转型计划,如何帮助相关国家向绿色低碳经济转型。

在"一带一路"全局层面:

- "一带一路"对全球生产、贸易与流通的影响是否足以帮助全球实现2050年CO_2减排的目标。

(3) 理解数字"一带一路"的作用

- 由于缺乏项目数据,加之专业人才和能力不足,部分项目所在国研究机构对"一带一路"及其影响的相关研究十分匮乏。

- 数字"一带一路"可以为共建"一带一路"国家打造一个平台，共享合作项目的相关数据。在第二届"一带一路"国际合作高峰论坛正式发布的"一带一路"生态环保大数据服务平台就是一次有益的尝试。这种发展得益于中国科技实力的崛起和数以万计科研人员与学者的努力，但在落实过程中需要推动中低收入国家的大学和研究机构的广泛参与。

- 创建"一带一路"研究网络基金，对获得"一带一路"利益相关方认可的独立联合研究团队进行资助。

- 数字"一带一路"可以促进各国之间理念、方法和实践的交流。

（4）实施有需求、可持续的项目

- 发展中国家希望能够与大型经济体展开合作，项目开发过程中容易忽略当地对可持续发展和落实2030年可持续发展议程的需求。

- 项目所在国在开发项目实施过程中可能没有严格的环境要求。

- 部分项目所在国对如何实现绿色合作与发展可能缺乏足够的认识。

- 需要增强意识，让各方认识到，投资根据当地情况、本土设计和景观制定的绿色解决方案与投资常规方案相比，能够产生同样甚至更大的效益。

- 需要建立合适的渠道，在与"一带一路"相关的社会环境和经济问题上与利益相关方进行沟通（本土或独立研究机构、当地企业、公民社会团体），以便他们的意见能够得到重视，并被纳入项目解决方案中。在巴基斯坦，一些中国企业与当地的国际非政府组织紧密合作，这一实践值得推广。

- 在斯里兰卡和巴基斯坦等国成功推动了"一带一路"环境友好项目的落地，可以为其他共建国家的项目起到示范作用。

（5）参与环境相关的具体项目

- 确定重点领域，识别目标清晰、可操作、可以产生环境效益的项目。

- 参与切实合作，分享各自在环境相关问题上的经验，并提供相应的技术和（或）资金援助。相关领域包括水治理、生态保护红线、国家公园管理、大规模森林修复、干旱地区修复等。

4.2 中国—马来西亚钦州产业园区

位于中国广西壮族自治区钦州市的中国-马来西亚钦州产业园区（以下简称中马钦州产业园）是"一带一路"绿色发展的典型案例之一。广西是"一带一路"重要节点和中国绿色发展优势地区。钦州市地处广西壮族自治区南部，是"一带"与"一路"的交汇点，是中国—东盟交往合作的最前沿地区。钦州市近年来高度重视生态文明建设和绿色发展，尤其注重中马钦州产业园的绿色发展。中马钦州产业园在国际产业园建设合作中的绿色理念、绿色制度、绿色规划和绿色措施等，对其他国家和地区共建绿色"一带一路"有着重要的参考价值。

中马钦州产业园是中国和马来西亚两国政府重点合作项目，是中国政府与外国政府合作共建的第三个国际园区。园区规划总面积 55 km^2，重点发展生物医药、电子信息、装备制造、新能源、新材料、现代服务业和东盟传统优势产业，致力于建设高端产业集聚区、产城融合示范区、科教和人才资源富集区、国际合作与自由贸易试验区。首期开发建设 15 km^2，其中启动区 7.87 km^2。园区以打造中国—东盟合作的示范区域——"中马智造城，共赢示范区"为发展目标，定位为：①先进制造基地。重点培育生物、多媒体等先进制造新兴产业聚焦区。②信息智慧走廊。借鉴马来西亚"多媒体超级走廊"（MSC）计划的成功经验，发展生物、多媒体技术产学研一体化的智慧园区。③文化生态新城。将中马钦州产业园的城市空间与周边的自然山体、主要水系（金鼓江）有机融会，创造绿色生态的山、水、城相融合的空间。④合作交流窗口。依托临近东盟的优势区位，将中马钦州产业园打造成服务中国—东盟自由贸易区的信息发布平台、贸易往来平台、项目展示及商务合作窗口。

4.2.1 中马钦州产业园绿色发展的主要举措

（1）制定实施包含绿色发展在内的园区管理条例

中国鲜有为园区专门立法的情况。而中马双方高度重视依法建园，由广西壮族自治区人民代表大会于 2017 年颁布实施《中国—马来西亚钦州产业园区条例》（以下简称《条例》）。《条例》是园区的基础性制度，其中对绿色发展做出了明确规定。

《条例》第四章第二十八条规定：产业园区应当坚持绿色发展理念，构建绿色产业体系和空间格局，完善生态保护设施和措施，大力发展生态产业，建设节能环保智慧园区，推进产业和城市一体化生态新城建设。

《条例》第四章第三十四条规定：产业园区应当建立和完善产业园区生态环保指标体系，推广产业园区能源循环利用、水资源综合利用和废弃物的减量化、无害化、资源化，控制重点污染物排放总量，促进低碳循环经济发展。

产业园区应当严格环境准入，完善环境保护基础设施建设，积极发展环保产业，禁止建设高能耗、高污染和高环境风险的产业项目，支持低能耗、低排放企业发展，推动企业实施清洁生产，保护和改善环境。

由此不难看出，中马钦州产业园突出强调了绿色发展理念，严格产业准入，严禁发展高能耗、高污染、高排放的产业。

（2）制定实施包括绿色发展在内的园区总体规划

制定实施《中国—马来西亚钦州产业园区总体规划》。该规划于2013年6月由广西壮族自治区政府正式批复，明确了"塑造品质、产城融合、创新机制、开放共赢、富民和谐、绿色生态"的园区建设总体目标和要求，其中"绿色生态"是主要目标和原则之一。

该规划于2018年1月进行修订。通过规划的制定和修订，产业及园区绿色发展的理念更加具体化、规范化，为园区绿色发展提供了坚实的规划指引。修订后的总体规划，更加突出了产业绿色化发展，重点对产业集群（板块）做了调整，由装备制造、电子信息、食品加工、材料及新材料、生物技术和现代服务业等六大产业，调整为电子信息、智能制造、生物医药、新能源、新材料、现代服务业和东盟传统优势产业等七大产业，其中剔除了能耗较高的装备制造业、污染较重的食品加工业。

另外，《中马钦州产业园区启动区控制性详细规划》进一步明确了"整体统筹、以人为本、特色构建、低碳和谐"的建设目标和原则，强调采用清洁生产技术、探索园区内生态链和生态网建设，最大限度地提高资源利用率，发展循环经济，注重园区自然景观保护和建设，并于2020年6月成功入选广西壮族自治区第三批绿色园区示范单位。

（3）以国家级绿色产业园区为建设目标

中国政府有关部门（如工业和信息化部等）高度重视绿色制造和绿色园区建设，相继发布了《工业绿色发展规划（2016—2020年）》《绿色制造标准体系建设指南》《工业节能与绿色标准化行动计划（2017—2019年）》等文件，鼓励和支持绿色园区建设。中马钦州产业园作为建设中的园区，建设伊始就把国家绿色园区作为目标，各项建设管理工作均以此为标准推进和完善。

（4）注重园区生态保护红线的划定与管控

园区管理委员会委托专业机构完成了《中马钦州产业园区基本生态控制线规划研究报告》。该报告对园区的生态红线进行了划定，明确了包括红树林在内的一级生态管制区，包括湿地在内的二级生态管制区。同时，该报告还明确了对各类生态管制区的管理措施，其中对于一级生态管制区，严格禁止任何破坏并强制修复红树林；对于二级生态管制区限制人类开发活动并注意加强生态保护与修复等。该报告还对绿色生态空间的比例与布局等进行了明确。

（5）钦州市政府在全市域开展包括绿色发展在内的园区系统评估

根据《钦州市开展园区区域性评估试点工作实施方案》，以中马钦州产业园为试点园区，对园区的水土保持、地质灾害危险性、压覆矿产、地震安全性、气候影响、文物保护等方面的情况进行评估[①]。

4.2.2 中马钦州产业园绿色发展的初步成效

一是园区大气、水环境质量保持基本平稳。园区特别重视发展分布式能源和太阳能发电，据当地生态环境部门监测，园区内空气质量与钦州市其他地区没有明显的差异。另外，园区水环境功能区水质达标率为100%，没有发生明显的水体污染事件。

二是园区主要绿色指标水平持续提升。特别是单位地区生产总值能耗，水耗，化学需氧量，二氧化硫、氮氧化物、氨氮排放量，工业固体废物综合利用率等园区主要绿色发展指标，均高于全国绿色园区平均水平。

三是园区道路绿色网络建设取得进展。园区林网建设取得阶段性进展，尤其

---

① 钦州市人民政府. 钦州市开展园区区域性评估试点工作实施方案[EB/OL]. [2019-01-08]. http：//www.qinzhou.gov.cn/phoneSite/zcjd/201901/t20190108_1977888.html.

是包括道路沿线、河流沿岸在内的园区绿色生态隔离网初步形成。同时，较为有效地保护了红树林、湿地等重要生态资源等。

四是入园产业和企业符合绿色环保要求。严格按照园区规划所确定的产业集群发展目标，强化对入园企业的绿色审核。截至 2018 年年底，落户园区的 350 家企业，主要包括金融服务业、现代物流业、文化产业等现代服务业；包括电子信息、现代装备制造在内的现代制造业；包括生物医药、纳米技术、云计算在内的战略性新兴产业等。

4.2.3　产业园区绿色发展的主要启示

一是明确坚持生态优先、绿色发展理念。生态文明建设已在中国普遍推进，其核心就是生态优先、绿色发展。生态优先，就是要以不突破生态保护红线为前提，以保护生态系统为目标；绿色发展，就是要推进资源节约、环境友好、生态保育。

二是将绿色发展全面纳入园区发展规划中。绿色发展，必须纳入产业发展、园区建设的规划体系中，以确保绿色发展落实到各产业门类、各空间单元。为此，应将包括资源节约、环境友好、生态保育的绿色发展，作为核心内容纳入园区建设和发展规划中。

三是将绿色发展相关内容纳入园区立法。园区建设与发展是一个长期、复杂的工程，需要强有力的法律保障，为此需要加强园区立法。为确保园区绿色发展，需要将绿色发展纳入园区立法之中，让绿色发展真正成为强制性遵守的原则和必须实现的目标。

四是必须严格把好入园项目的绿色审查关。产业园必须重视入园项目的绿色化，严格禁止资源消耗大、环境污染重、生态占用多的项目或企业入园。要严格把好入园项目的绿色审查关，必要时建立项目绿色审查制度。

五是建立园区绿色发展动态评估机制。建立产业园区绿色发展的动态评估机制和园区项目退出机制，及时发现非绿色的项目和企业，并对其进行警告、限期整改，对于逾期不改或虽整改但未达到要求的，可以予以劝离。

5 推动绿色"一带一路"建设政策建议

2030年可持续发展议程对于全球各国来说都是一个具有深远意义的可持续发展框架，也是各国未来努力发展和希望实现的目标与共识。而绿色"一带一路"建设对共建国家的意义和效益都有所不同，只有在各方认识到绿色"一带一路"对其长远可持续发展有积极贡献基础上才能推动共建行动，因此，首先要在涵盖包容性、协调性、一致性、能力建设等"一带一路"建设主要原则的基础上，加强各国可持续发展目标落实计划与绿色"一带一路"建设的战略协调（图5-1）。其次，从绿色"一带一路"内涵来看，主要是将绿色发展理念融入"五通"，尽可能在"一带一路"建设过程中减少对生态环境的影响，特别是"设施联通"和"贸易畅通"两个重点合作领域的绿色化，而"绿色政策沟通""绿色资金融通""绿色民心相通"将提供政策、资金上的支持，营造外部友好的氛围。

图 5-1 绿色"一带一路"建设路径

5.1 积极参与全球环境治理与气候治理，将"一带一路"打造成全球生态文明和绿色命运共同体的重要载体

通过绿色"一带一路"构建绿色发展国际合作伙伴关系与网络。 保护全球生态环境，需要世界各国同舟共济、携手同行，构筑尊崇自然、绿色发展的生态

体系。建设生态文明既是中国作为最大发展中国家在可持续发展方面的有效实践，也是为全球环境治理提供的中国方案和中国贡献。"一带一路"建设应秉持生态文明理念，为构建人类绿色共同体提供中国智慧，为后发国家避免传统发展路径依赖和锁定效应提供可资借鉴的示范模式和实践经验，帮助更多地区采纳并落实可持续发展行动。

通过"一带一路"绿色发展国际联盟建设，开展"一带一路"生态环保合作。"一带一路"绿色发展国际联盟定位于建设开放、包容、自愿的国际合作伙伴关系，以联盟为依托，打造政策对话和沟通平台，宣传和分享生态文明和绿色发展的理念与实践，推动将绿色发展理念融入"一带一路""五通"，进一步凝聚国际共识。以联盟为依托，打造环境知识和信息共享平台，加强同国际智库间的联系，开展联合研究，共同促进共建"一带一路"国家落实联合国2030年可持续发展议程。

通过绿色"一带一路"积极参与全球环境治理与气候治理进程，共同开展全球生态文明建设。通过绿色"一带一路"推动共建国家积极参与全球环境治理体系改革和建设，增强同舟共济、权责共担的命运共同体意识，协调各方立场，推动商定公平公正的全球环境治理规则，提升全球环境治理水平，为全球层面落实2030年可持续发展议程、完善全球环境治理体系提供新思路、新方案。

通过绿色"一带一路"传播生态文明理念，在共建国家推动形成生态文明建设共识。生态文明和可持续发展都产生于全球经济治理格局深度调整的时代背景之下，都旨在建设人类的绿色家园，尽管二者的角度和立场，以及要实现的最终目标表述有所不同，但从内容上看，相互间存在许多契合点。应促进绿色"一带一路"建设与联合国可持续发展目标对接，加强中国生态文明与共建国家可持续发展理念互学互鉴、相互理解和支持。

5.2 建立"一带一路"建设对接机制，以政策、规划、标准和技术对接促进战略对接并落地

5.2.1 推进绿色"一带一路"建设战略对接

把绿色"一带一路"作为中国与有关国家和国际组织签署合作共建"一带

一路"谅解备忘录的重要内容**。在正式签署的谅解备忘录中，明确写入双方致力于秉持生态文明和绿色发展理念，合作共建绿色"一带一路"、促进"一带一路"建设与联合国2030年可持续发展议程对接有关内容。立足现有的我国与共建国家以及国际组织之间签署的双（多）边合作战略协议，与合作方成立生态环境常设性工作组，具体负责双方绿色发展战略和规划衔接工作。

依托交流平台和双（多）边合作机制开展战略对接。一方面，在中国—东盟环境合作论坛、欧亚经济论坛、中阿环境合作论坛等对话机制中固定设置绿色"一带一路"建设议题。探索开展绿色"一带一路"项目建设和融资指南、技术标准等领域合作，以标准规则建设促进战略对接。把绿色"一带一路"合作项目作为未来几年中俄、中哈、中欧以及中国—东盟、上合组织、金砖国家等双多边合作机制的重点领域，以项目合作促进战略对接。

促进与东道国生态环境保护政策对接。对"一带一路"项目进行监管，以便更全面地掌握项目对当地和全球环境的影响；落实国际合作伙伴关系，为建立全面系统的监测体系提供支持。合作伙伴关系应当参与"一带一路"政策沟通，与利益相关方共同探讨明确界定绿色项目与非绿色项目的标准。

对接"一带一路"与南南合作。"一带一路"倡议与南南合作有诸多契合之处，在南南合作框架下推进绿色"一带一路"，强化顶层设计，将其打造成中国与发展中国家新南南合作的典范，加大援助、产业合作、能力建设等多种形式的合作。组建国际研究小组，帮助发展中国家更好地推动绿色发展和可持续发展进程。

5.2.2 强化生态环境合作融入"一带一路"建设全过程

充实强化"一带一路"建设相关规划中生态环境合作内容。研究制定"一带一路"建设合作规划编制指南，对其中生态环境保护和绿色发展合作内容做出强制要求，深化细化相关综合性规划和各专项规划中涉及生态环境保护相关章节的内容，在未来制定相关文件时把绿色"一带一路"建设作为重要内容。

与合作项目较多的共建国家联合编制生态环境保护规划。与已编制基础设施互联互通和国际产能合作规划、生态环境问题比较突出的共建国家，联合编制生态环境保护和绿色发展规划，并在未来开展新合作时，同步编制相关规划，确保绿色发展理念全方位融入"一带一路"合作中。

推动生态环境标准合作与应用。 加强中国与共建"一带一路"国家绿色基础设施标准的协调与联通，开展联合研究，开发在"一带一路"区域认可的绿色交通、绿色建筑、绿色能源等领域的国际标准。利用"一带一路"环境技术交流和转移中心（深圳）以及环保技术和产业合作示范基地，支持企业与共建国家开展生态环保技术合作，与相关行业协会共同制定并发布双方都认可的生态环保行业标准。

促进"一带一路"项目之间的技术对标。 "一带一路"项目多为跨国项目，并且由于技术方面的限制，在开发过程中面临一系列复杂问题。应成立特别委员会负责制定统一的"一带一路"重要技术标准，加强标准对接；对于有生态环境积极影响的重点绿色产业，应简化公开招标和国际竞标流程。

5.3 构建绿色"一带一路"源头预防机制，以绿色金融、生态环境影响评价等机制引导绿色投资

5.3.1 强化绿色金融支持绿色"一带一路"建设

在国际层面采用绿色金融工具推动"一带一路"建设绿色化。 一是研究"一带一路"绿色投融资原则。在充分吸收国际原则、标准的基础上，研究制定"一带一路"绿色投融资原则和指引，促进绿色投融资理念的形成。二是成立共建"一带一路"国家参与的"一带一路"绿色投融资担保机构，为绿色项目及节能减排等投融资活动提供必要的担保，应对环境气候风险，撬动商业资金进入绿色投资领域。三是在"一带一路"生态环保大数据服务平台建设环境与社会信息数据库，为"一带一路"合作项目的相关投资人、贷款人、融资人、业主等提供信息服务。四是推动为"一带一路"项目提供金融服务的机构开展环境信息披露。推动金融机构尽快重视环境气候风险并支持绿色发展，改善企业绿色绩效表现，从而促进经济可持续发展。

鼓励共建国家政府将绿色金融作为绿色转型的重要工具。 助力提升共建国家绿色金融能力，积极分享绿色金融相关经验。一是培育绿色投融资市场需求，鼓励绿色产业、绿色技术和绿色项目发展。二是完善金融监管政策，鼓励金融机

构积极支持绿色产业、技术、企业和项目，引导和鼓励金融机构建立绿色投融资机制。三是积极培育有环境社会责任的投资人。

发挥金融机构中介作用改善企业客户环境表现。一是促进金融机构建立清晰的绿色金融发展战略。培育绿色理念、价值观，建立绿色金融组织架构，积极拓展绿色金融市场，有效防范环境和气候风险。二是建立和完善境外业务绿色金融政策制度。创新绿色金融产品，提升绿色金融服务能力。三是建立环境与社会风险评估方法。在借鉴国际金融公司（IFC）绩效标准、赤道原则等国际通行标准的基础上，分析"一带一路"沿线环境、社会具体情况，建立合理的评估方法和工具。四是环境社会风险的全流程管理。将环境与社会风险管理纳入信贷和投资管理的前、中、后管理流程，明确岗位职责。五是实施环境信息披露制度。建立环境信息披露制度和框架，提升披露能力。六是建立环境与社会风险应对机制。

5.3.2 建立"一带一路"建设项目生态环境管理机制

建立"一带一路"项目分级分类管理机制。建立"一带一路"建设投资项目生态环境管理数据库，将生态环境影响因素纳入"一带一路"建设项目风险评估体系；同时从经济风险、政治风险、社会风险、文化风险、生态环境风险等方面对"一带一路"建设项目进行风险评估；从生态安全、环境污染等维度明确投资项目的生态环境影响程度并评估项目环境收益。并将以上这些作为开发性金融和政策性金融支持的重要标准。

开发"一带一路"建设项目环境管理工具。开发投资项目环境、气候和社会风险识别、评估、监测及管理工具，研究包含政策、法规、数据信息等在内的投资咨询服务工具，依托"一带一路"生态环保大数据服务平台，建立"一带一路"重点投资国别的生态环境信息系统和评估工具，开发配套技术支持工具，加强公共环境数据可得性。对生态敏感脆弱地区的生态环境风险进行综合全面、深入细致的评估，建立风险清单和管控措施清单，并要求投资项目落实有关清单任务和要求。在评价范围上，结合国际关注热点，基于自愿原则，鼓励将生态健康和气候变化等因素纳入环评范围中。

完善"一带一路"建设项目环境管理平台和流程。搭建环境保护评估咨询服务平台、政企银投融资信息共享平台等，针对"一带一路"建设项目相关信息披露进行独立审议以确保项目环境与社会保障措施的实施，保护项目相关方利益。

鼓励利益相关方参与项目环境管理。 在环评过程中，鼓励利益相关方全面、有效参与，严格按照东道国相关法律法规实施，在科学的基础上客观准确评估项目的生态环境影响。

5.4 构建"一带一路"项目管理机制，推动企业落实绿色发展实践

加强绿色供应链管理。 建议推动中国优势产业对接并融入全球供应链体系，联合"一带一路"国家相关部门、机构和企业共同打造区域绿色供应链体系，充分发挥各国产业优势和市场优势，开展更大范围、更高水平、更深层次的国际合作。充分利用"一带一路"绿色供应链合作平台，支持和鼓励企业积极开展对外贸易与投资合作，促进供应链上的企业进行绿色创新。开展绿色供应链管理试点示范工作，制定绿色供应链环境管理政策工具，从生产、流通、消费的全产业链角度推动绿色发展进程。构建绿色供应链绩效评价指标体系，评价企业绿色供应链管理绩效，提升企业可持续发展社会责任感。

探索设立"一带一路"绿色发展基金。 加大资金投入力度，保障绿色"一带一路"相关工作的落实开展。推动设立专门的资源开发和环境保护基金，重点支持共建线上国家生态环保基础设施、能力建设和绿色产业发展项目。同时发挥国家开发银行、进出口银行等政策性金融机构的引导作用，引导、带动各方资金投入绿色发展基金，共同为绿色"一带一路"建设提供支持。

发展绿色价值链。 一是加强发达国家绿色技术溢出的捕获与推广，强化中国捕获发达国家绿色技术溢出的能力，加强对捕获的绿色技术在共建"一带一路"国家的示范与推广。二是在国际产能合作和科技工业园区建设中推进绿色价值链，一方面，对于钢铁、化工、油气开采等污染密集型行业，应大力推进行业环境规制的互通互认，促进节能减排，降低环境负外部性；另一方面，对于核能、水能、风能、太阳能等清洁能源以及其他环境友好型行业，应加大扶持力度，保障项目培育和实施，促进经济、社会、环境的可持续协调发展。三是以绿色标准和标识引领绿色价值链发展，为产品全生命周期关键环节以及价值链主要相关方

建立绿色标准，强化绿色价值链中投入与产出的环境友好属性，对产品涉及的从原材料到最终产品再到淘汰处理的全过程建立绿色标准并构建评价与认证机制。

促进环境产品与服务贸易便利化。促进绿色产品与服务的贸易便利化将为共建"一带一路"国家带来巨大的环境效益。建议提高环境产品与服务市场的开放水平，发展绿色产业，鼓励扩大大气污染治理、水污染防治、固体废物管理及处置技术和服务等环境产品和服务进行进出口贸易。为绿色产品与非绿色产品制定差别化政策，对产品贸易与投资实行分类指导与管理，如给予绿色产品降低关税的优惠、通关便利化待遇、将绿色产品纳入产业投资鼓励目录、提供绿色金融服务等，争取世界银行、亚洲开发银行等国际金融机构对绿色项目的优惠贷款。强化绿色产品标识与认证体系及国家间的绿色产品互认，推动"一带一路"沿线各国政府采购清单纳入更多的环境标志产品。

5.5 通过民心相通加强绿色"一带一路"建设，强化人员交流与能力建设

将绿色丝路使者计划打造为环保能力建设的旗舰项目。绿色丝路使者计划是中国开展南南合作，促进区域可持续和绿色发展的重要平台。建议将绿色丝路使者计划打造成在共建"一带一路"国家公众环境意识提高以及环保能力建设领域的旗舰项目，立足于"政策沟通""民心相通"，通过开展环境管理人员和专业技术人员培训、政策内容指导活动等形式加强在环境治理、污染防治、绿色经济等领域的合作交流，分享中国生态文明和绿色发展的理念与实践；推动地方政府参与绿色丝路使者计划，借助中国—东盟环保技术和产业交流示范基地、"一带一路"环境技术交流与转移中心（深圳）等平台，引导环保企业有序地"走出去"。

支持和推动中国与共建国家环保社会组织交流合作。构建政府引导、企业支持、社会参与、行业互助"四位一体"的支持网络，明确政府部门推动主体，出台政策或指引性文件明确中资企业的海外环境责任，引导环保社会组织建立合作网络。引导形成多元化的资金机制，加大政府购买环保社会组织服务的力度，设立支持环保社会组织"走出去"的专项合作资金。完善环保社会组织参与机制，建立协商与决策参与机制，建立环保社会组织参与的国际交流事项清单。

推动社会性别主流化，提升女性领导力。提升政策制定者和妇女群体的社

会性别意识，推动社会性别意识纳入绿色"一带一路"政策制定与项目实施。在"一带一路"项目建设过程中落实性别主流化最佳实践，借助绿色丝路使者计划，组织共建"一带一路"国家生态环保领域女性官员、专家学者、青年学者等开展"提升女性绿色领导力"的专题项目培训，并与"一带一路"合作伙伴分享实现性别主流化的方法与经验。

Part 2 英文部分

Green BRI and 2030 Agenda for Sustainable Development
——Accelerating Effective Alignment and Synergy between BRI and 2030 Agenda to Promote Global Ecological Civilization

绿色"一带一路"与 2030 年可持续发展议程
——有效对接与协同增效　共谋全球生态文明建设

Executive Summary

2030 Agenda for Sustainable Development and *Paris Agreement* approved on the 2015 UN Summit marked the start of a new era of sustainable development. Against the backdrop, China proposed the construction of green silk road. On the Belt and Road Forum for International Cooperation held in May 2017, Chinese President Xi Jinping proposed that "We should pursue the new vision of green development and a way of life and work that is green, low-carbon, circular and sustainable. Efforts should be made to strengthen cooperation in ecological and environmental protection and build a sound ecosystem so as to realize the goals set by the *2030 Agenda for Sustainable Development*". In April and May 2017, The Ministry of Environmental Protection of the People's Republic of China (now the Ministry of Ecological Environment) and other related ministries launched *Guideline of Promoting the Construction of the Green Belt and Road Initiative* and *the Plan for Eco-environmental Protection Cooperation under Belt and Road Initiative*, which identified the general direction and cooperation framework of ecological and environmental protection under the Belt and Road Initiative in the coming decade.

Building a green Silk Road is in line with the global trend of green development and the practical need of developing countries for green development and environmental protection. It also coincides with the principles of the *2030 Agenda for Sustainable Development*. Generally speaking, most countries along the Belt and Road are developing countries and emerging economies with the need to balance development and conservation in the process of industrialization and urbanization. The construction of green Belt and Road will facilitate the cooperation among countries along the Belt and Road on environmental protection, promote the improvement in environmental management capacity for countries along the Belt and Road and push forward the implementation of the 2030 Sustainable Development

Goals in the region.

In the past 6 years, the Belt and Road Initiative has been upholding the green development principles, emphasizing alignment with the *2030 Agenda for Sustainable Development*, promoting the green and low-carbon construction, operation and management of infrastructure and enhancing cooperation in ecological and environmental protection, biodiversity conservation and climate change mitigation. These efforts have created new impetus for the implementation of 2030 Sustainable Development Goals and delivered new opportunities for green development in related countries. First, Green Belt and Road will promote policy communication in ecological and environmental protection and alignment of Green Belt and Road and SDG on policy level; second, Green Belt and Road will construct an integrated ecological and environmental risk prevention and governance system to prevent environmental risks caused by infrastructure construction; third, Green Belt and Road will promote green and unimpeded trade, improve the efficiency of production and consumption, promote green financial integration and guide investment to clean energy and other green industries; fourth, Green Belt and Road will enhance people-to-people bond to promote ecological and environmental protection capacity building in developing countries.

The Belt and Road Initiative is a large and ambitious program with both opportunities and challenges. In terms of green-oriented philosophy, many countries are under-developed and they never had the chance before to get steadily in touch with such concepts. In terms of policy and monitoring assessment, BRI projects are mostly very complex and transnational, they involve different standards and procedures to plan, design, construct, operate and assess projects. In terms of green finance and green investments, there is a lack of policy guidance; in terms of projects, BRI projects are mostly large infrastructure projects that create both opportunities and risks.

The Belt and Road is a road to green development that calls for the efforts of all sides. A green Silk Road requires the establishment and implementation of the strong consciousness that lucid waters and lush mountains are invaluable assets and joint efforts in promoting ecological civilization construction worldwide. A Green Belt and Road will provide more green public goods to BRI participating countries and regions and effectively promote the implementation of the *2030 Agenda for Sustainable Development*. It is believed that with concerted efforts of Chinese and foreign partners, fruitful outcomes will be achieved in greening the Belt and Road.

The research concluded that constructing Green Belt and Road need to build consensus among all parties and explore to establish long-term mechanism from perspectives of strategy, planning, evaluation and financial support. The establishment of long-term mechanism of green BRI should start from strategic alignment, planning embedding, project evaluation, financial support, technology promotion etc. In addition, the construction of a more inclusive global green value chain is also a key element for green BRI.

This book analyzes the ecological and environmental protection situation of the BRI projects in Pakistan and Sri Lanka, as well as the green development cases of China-Malaysia Qinzhou Industrial Park, and proposes green development principles for the BRI projects, especially infrastructure construction projects, and clarifies the ways to align the green BRI with the 2030 Agenda for Sustainable Development from the implementation level.

Based on the theories and practical cases, this book puts forward policy recommendations to promote green Belt and Road: 1) Play an active role in global environmental governance and climate governance, transforming the Belt and Road Initiative into an important instrument for global ecological civilization construction and building a green community of common destiny. 2) Promote strategic alignment in the development of the green Belt and Road with connection of policies, planning, standards and technologies. 3) Design safeguard mechanisms for green Belt and Road from its source, and guiding green investment with mechanisms of green finance and ecological impact assessment. 4) Establish a mechanism for BRI project management and promote the enterprises to adopt practices on green development. 5) Enhance people-to-people exchanges and capacity building to boost green BRI.

1 Green Belt and Road and 2030 Agenda for Sustainable Development

1.1 The proposal and development of the Belt and Road Initiative

Since the outbreak of the global financial crisis in 2008, the world economy has been sluggish. Trade growth has been slow and instability persists. There is an urgent need of global economy for new growth engines and new cycles. The huge demand for infrastructure and industrial development in developing countries, emerging economies included, is expected to serve as the new momentum for economic growth.

It was against this backdrop that in 2013, Chinese President Xi Jinping proposed in Kazakhstan and Indonesia to build the Silk Road Economic Belt and the 21st-Century Maritime Silk Road, namely the Belt and Road Initiative. In March 2015, the Chinese government issued *Vision and Actions on Jointly Building Silk Road Economic Belt and 21st-Century Maritime Silk Road*[①] (hereinafter referred to as *Vision and Action*), which proposes the top-down design framework for jointly building the Belt and Road, including objectives and vision, principles and future potential and directions. According to the *Vision and Action*, the BRI is aimed at promoting orderly and free flow of economic factors, highly efficient allocation of resources and deep integration of markets. It encourages the Belt and Road participating countries to achieve economic policy coordination and carry out broader and more in-depth regional cooperation of higher standards; and advocates jointly creating an open, inclusive and balanced regional economic cooperation architecture.

The BRI takes achieving shared growth through discussion and collaboration as its fundamental principle. The core lies in encouraging the BRI participating countries

---

[①] National Development and Reform Commission, Ministry of Foreign Affairs, Ministry of Commerce of the People's Republic of China. Vision and Actions on Jointly Building Silk Road Economic Belt and 21st-Century Maritime Silk Road[EB/OL]. [2015-03-28]. http://www.mofcom.gov.cn/article/resume/n/201504/20150400929655.shtml.

to align and coordinate development strategies, build consensus to the maximum extent, and leverage their respective comparative advantages, so as to share the achievements of the initiative and the long-term dividends. Priorities for the initiative include policy coordination, facilities connectivity, unimpeded trade, financial integration and people-to-people bond.

Since being proposed, the Belt and Road Initiative has been well-received in more and more countries. Now, it has become a "Chinese solution" for participation in global openness and cooperation, improving global environmental governance system, promoting shared development and prosperity around the world and building a community of shared destiny for mankind. In May 2017, the Belt and Road Forum for International Cooperation (BRF) was held. 29 heads of states and representatives from more than 130 countries and 70 international organizations reaffirmed the BRI core principles of consultation, contribution and shared benefits. In April 2019, the Second Belt and Road Forum for International Cooperation was successfully held. 38 heads of states, and 40 leaders of international organizations including Secretary General of the United Nations and IMF Chief attended the Leaders' Roundtable of the 2nd BRF. Over 6,000 international guests from 150 countries and 92 international organizations attended the 2nd BRF, which provides a platform for participants to exchange opinions in-depth with each other on jointly implementing the BRI. It is widely accepted that the Belt and Road marks a road of opportunities with consensus achieved on realizing high-quality development on the Belt and Road, and fruitful outcomes have been achieved.

In the past 6 years, the Belt and Road Initiative has developed from a concept and vision to concrete actions, and entered the phase calling for full implementation and outcomes delivered. From 2013 to 2018, the total volume of trade in goods between China and countries along the Belt and Road exceeded 6 trillion USD. China has invested more than 20 billion USD in developing overseas economic and trade cooperation zones, creating hundreds of thousands of jobs and several billion USD of tax revenue for the local area. A series of cooperation projects have achieved concrete progress. China-Pakistan Economic Corridor (CPEC) is being developed according to schedule, China-Laos Railway, China-Thailand Railway and Hungary-Serbia Railway is under construction, parts of Jakarta-Bandung High Speed Railway have been put into construction, Gwadar Port is ready to be put into full operation. China Railway Express to Europe has connected 115 cities in 17 countries on the Asia-Europe continent, with a total of more than 21,000 trains operated.

1.2 Progress of building a green Belt and Road

1.2.1 The ecological and environmental condition of key areas along the Belt and Road

The Belt and Road participating countries along the Belt and Road have diverse environmental and climate conditions with shared ecological and environmental challenges. Most of them are developing countries in Southeast Asia, South Asia, West Asia, ASEAN and North Africa. With excessive population growth and rapid industrial expansion, soaring resource consumption and pollution discharge is putting increasing pressure on the environment.

(1) The overall ecological environment is sensitive

On the whole, the BRI participating countries are facing various ecological and environmental problems, and the ecological environment is generally more sensitive.

China-Pakistan Economic Corridor faces the challenge of drought in Xinjiang. Northern Pakistan is located at the intersection of major mountain ranges, including the Har-Goolun Range, the Himalayas and the Hindu Kush Mountains. As a biodiversity hotspot, the region also poses huge challenges to realizing connectivity in China-Pakistan Economic Corridor. Southwestern Pakistan suffers from severe drought and water shortage, while Southeastern Pakistan is faced with air, water and soil pollution, combined with seasonal water shortage.

China-Mongolia-Russia Economic Corridor covers large patches of permafrost in Southwest Russia which is located in the frigid and sub-frigid zones. Already being challenged with deforestation, the region has a fragile eco-system that is hard to recovery once being damaged. The Mongolian plateau is faced with a lack of surface water, low forest coverage and a fragile desert eco-system.

Along the China-Indochina Peninsula Economic Corridor, Lancang-Mekong countries such as Cambodia, Thailand, the Laos and Vietnam have rich water resources, high forest coverage rate and diverse flora species. Disputes over water resources and water pollution are the most pressing issues in the region. The Malay Peninsula is also faced with the problem of environmental pollution.

The regions covered by China-Central Asia-West Asia Economic Corridor are mostly mountains, oasis and deserts, with water shortage and transboundary disputes over water resources as the main challenges. Central Asian nations are also faced with the challenge of desertification and nuclear pollution.

Bangladesh-China-India-Myanmar Economic Corridor faces the challenge of segmented ecological landscape and decrease in biodiversity as a result of deforestation in mountainous areas and highlands. The low-mountain plain area are faced with the major problem of non-point source agricultural pollution and cultivated land degradation. Coastal areas in the Bay of Bengal are suffering from problems such as the damage of mangrove forests and sea level rise caused by climate change.

New Eurasian Land Bridge Economic Corridor faces the challenge of drought and desertification in Western areas. The section of the Bridge in Central Asia has draught and a fragile desert eco-system as major eco-environmental challenges, while the section in Russia and Europe is dotted with natural reserves of all kinds.

(2) Water pollution disrupts regional economic development and social stability

Located in the hinterland of Eurasian continent, Central Asia has a dry climate with desert and grassland as the main landform. As a result, water pollution and shortage is the primary environmental challenge in the region, which is reflected in the following aspects: 1) lakes are shrinking with worsening water quality; 2) river water volume keeps decreasing with the rivers becoming shorter or disappearing and water quality continues to deteriorate; 3) underground water levels continues to drop with deteriorating water quality; 4) areas affected by soil alkalization continues to expand; 5) deserts expand and oasis shrinks, resulting in more frequent sandstorms; 6) natural vegetation decreases and degrades. These changes are most obvious in the drainage area of the Aral Sea. The deterioration of the quality of water leads to declining birth rate and raising infant mortality rate, forcing local residents to move. Water Resources become one of the major factor that hindering the social and economic development of Central Asia.

South Asian countries are plagued by severe water pollution. For example, water bodies in many Indian cities are being polluted by domestic sewage, industrial waste water, chemical substance and solid waste. Water pollution, flood and draught has become three major water-related disasters faced by India, affecting sustainable development in every aspect of the Indian society. Without enough fund in water resource conservation, India witnesses direct discharge of industrial waste water into rivers and lakes, causing pollution of underground water sources, with the content of all chemical substances exceeding international standards. The content of lead in underground water bodies in India is more than 20 times higher than that in industrialized countries with improved waste water treatment facilities. Besides, the direct discharge of untreated domestic sewage also contributes to water pollution, posing health risks to the general public. The Ganges River, the major river in Northern India, has been recognized as one of the most seriously polluted rivers in the world. The use of polluted underground water for drinking and cooking has caused health issues among local residents. At the same time, 12 kinds of soft drinks sold on the market in India have been tested with exceedingly high content of harmful residue. The content of pesticide residue tested in some soft drinks is 10~70 times higher than European standard.

Middle East countries are suffering from water shortage with most of the population living in coastal areas or drainage areas of major rivers. In the past 2 decades, massive industrialization and limited infrastructure and economic development has contributed to further environmental deterioration in the Middle East, especially in some oil producing countries. Chemical pollutants such as sulphate and nitrate have been damaging the environment in oil producing countries such as Kuwait, Qatar, Saudi Arabia, Iran and Ira, threatening the environmental well-being of other countries in the same region. The Iran-Iraq war in the 1980s and the Gulf War in the early 1990s left the Gulf Region with a series of environmental issues with long-lasting effects. Oil drilling facilities and oil tanks were damaged in the war, causing massive oil leaks, threatening the existence of marine species in the Gulf Region.

(3) Air pollution is a prominent issue in countries along the Belt and Road

The level of air pollution in BRI countries is lower than the global average. However, certain regions suffer from severe air pollution.

South Asia, West Asia and North Africa suffer from particulate matter pollution ($PM_{2.5}$). Among the 65 BRI countries, 22 see relatively high $PM_{2.5}$ density, in which 11 are West Asian and North African countries and 6 are South Asian countries. Delhi, Lahore and Dhaka are among the most severely air-polluted cities in the world. Besides poor atmospheric diffusion caused by geographic limitations, the over-reliance on the oil and chemical industries is also a major factor contributing to the high $PM_{2.5}$ concentrations.

The air pollution in Northeast Asia and Southeast Asia is also severe. Ulaanbaatar, the capital city of Mongolia, provides a vivid example of air pollution issues in Mongolia and Russia. Rapid urbanization brought with it deterioration of the air quality in cities. Air pollutants emitted by domestic heating and cooking and the exhaust from vehicles, as well as the increasing emission from power generation as a result of increasing power demand from industries and construction, together contribute to the worsening air quality. In 2011, Ulaanbaatar was listed as one of the top 5 cities in the world with the worst air quality by WHO in its World Air Quality Report. In Southeast Asia, forest fires, industrial pollution and exhaust from vehicles have caused decreases of air quality, with the worst in Hanoi in Vietnam, Bangkok in

Thailand and Indonesia. It also needs to be noted that Eastern Europe is also faced with severe air pollution as a result of the development of heavy industries and power plants.

1.2.2 SDGs index of countries along the Belt and Road

(1) Environmental targets are important components of the 2030 SDGs

Since the 1960s, the international community has been exploring for a pathway to balance the three dimensions of sustainable development: the economic, social and environmental. In 2000, at the Millennium Summit of the United Nations, leaders of 189 countries adopted *the United Nations Millennium Declaration* to eradicate poverty, hunger, illiteracy and gender discrimination, reduce the spread of diseases and prevent environmental deterioration. The Declaration contained a statement of values, principles, and objectives for the international agenda for the 21st century to be achieved by 2015 (Millennium Development Gals, MDGs). Under the guidance of MDGs, the world has made positive progress in economic development, improvement of people's well-being and the eradication of hunger and poverty. In June 2012, at the Rio+20 Summit, global leaders proposed the development of an agenda that builds upon the achievements of the Millennium Development Goals with sustainable development as the core. *Transforming our World:* the *2030 Agenda for Sustainable Development* (*2030 Agenda* for short) was adopted at the United Nations Sustainable Development Summit on 25 September 2015. It announced Sustainable Development Goals (SDGs) that would stimulate action over the next fifteen years in areas of critical importance for humanity and the planet.

Sustainable Development Goals (SDGs) are core components of the *2030 Agenda*. The *2030 Agenda* is composed of four parts[①]: Declaration; Sustainable Development Goals and Targets; Means of Implementation and the Global Partnership; and Follow-up and Review. The *2030 Agenda*, with unprecedented scale

---

[①] Ministry of Foreign Affairs of the People's Republic of China. Transforming Our World: the 2030 Agenda for Sustainable Development[EB/OL]. [2016-01-13]. http://www.fmprc.gov.cn/web/ziliao_674904/zt_674979/dnzt_674981/ xzxzt/xpjdmgjxgsfw_684149/zl/t1331382.shtml.

and ambition, requires a revitalized global partnership to ensure its implementation. As the core of the *2030 Agenda*, the 17 Sustainable Development Goals (SDGs) and 169 targets are integrated and indivisible and balance the three dimensions of sustainable development: the economic, social and environmental. The Goals and targets are to be realized by 2030 or 2020. They are aimed at fundamentally changing the traditional view on development that focuses only on economic growth, providing guidance to policy-making and fund allocation for countries in the next 15 years, and stimulating action in areas of critical importance for humanity and the planet to end poverty and hunger, protect the planet and ensure that all human beings can enjoy prosperity. The first 16 SDGs define the goals and targets that the international community needs to achieve, and SGD17 (Global Partnership for Sustainable Development) emphasizes the importance of international cooperation in promoting the implementation of the *2030 Agenda*.

Environmental targets are important components of the 2030 SDGs. The *2030 Agenda* emphasizes challenges brought by resource and environment issues on human existence and livelihood. Environmental targets are almost prevalent in all goals and index, covering every aspect of ecological and environmental protection. After summarizing sustainable development goals and index and analyzing those relevant to environmental sustainability, it is found that 52.9% of the 17 goals and 14.2% of all targets are related to ecological and environmental protection. Some of the environmental targets are independently defined, and some are integrated into other development goals targets. To be more specific, the Goals related to ecological and environmental protection include: Clean Water and Sanitation (SDG6), Affordable and Clean Energy (SDG7), Sustainable Cities and Communities (SDG11), Responsible Consumption and Production (SDG12), Climate Action (SDG13), Life Below Water (SDG14), Life on Land (SDG15) and Partnerships for the Goals (SDG17) (Table 1-1). Other environment-relevant targets touch upon the issue of chemical substance pollution prevention and control, air quality and improvement of soil pollution.

Table 1-1 Environment-related goals and targets proposed by the *2030 Agenda*

Goals		Environment-related issues	Number of targets
Goal 6	Ensure access to water and sanitation for all	Clean water and sanitation	8
Goal 7	Ensure access to affordable, reliable, sustainable and modern energy	Sustainable modern energy	5
Goal 9	Building resilient infrastructure, promote sustainable industrialization and foster innovation	Sustainable industrialization	8
Goal 11	Make cities inclusive, safe, resilient and sustainable	Sustainable cities	10
Goal 12	Ensure sustainable consumption and production patterns	Sustainable consumption and production	11
Goal 13	Take urgent action to combat climate change and its impacts	Climate change	5
Goal 14	Conserve and sustainably use the oceans, seas and marine resources	Oceans and marine resources	10
Goal 15	Sustainably manage forests, combat desertification, halt and reverse land degradation, halt biodiversity loss	Land bio-system, forest, desertification, land degradation, biodiversity	12
Goal 17	Revitalize the global partnership for sustainable development	Sustainable global partnership	19

(2) Environmental sustainability goals and index in 2018 *SDG Index & Dashboards Report*

The SDG Index and Dashboards assess where each country stands with regard to achieving the Sustainable Development Goals (SDGs). Sound metrics and data are critical for turning the SDGs into practical tools for problem solving. Therefore, Sustainable Development Solutions Network (SDSN), in collaboration with other organizations, proposed the SDG Index and SDG Dashboards as the first worldwide study to assess where each country stands with regard to achieving the Sustainable Development Goals. Instead of being an official monitoring tool, it is a complement to the official SDG indicators and voluntary country-led review processes which uses OECD (2008) country specific construction comprehensive index to propose critical hypothesis on the basis of bridging the gap of related data with United Nations SDGs index and other reliable data to work out a set of measurement standards on state level. Governments and civil society alike can utilize the Sustainable Development Report to identify priorities for action, understand key implementation challenges, track progress, ensure accountability, and identify gaps that must be closed in order to achieve the SDGs by 2030.

The SDG Index and Dashboards Report is co-produced every year since 2015 by the Sustainable Development Solutions Network (SDSN) and Bertelsmann Foundation. In 2015, SDSN and Bertelsmann Foundation published a report titled *Sustainable Development Goals: Are the Rich Countries Ready*? Which described where the 34 OECD countries were in terms of the implementation of SDGs. The SDG Index and Dashboards Report has been published by SDSN and Bertelsmann Foundation on an annual basis since 2016. *The SDG Index and Dashboards Report 2017—Global Responsibilities: International Spillovers in Achieving the Goals* was launched by SDSN and Bertelsmann Foundation in July 2017. The Report ranked different country based on how leaders can deliver on their promise for each of the 17 goals and showed the general implementation of the 17 goals through color coding and present the final result as SDG Dashboards. A separate report for each country on their fulfillment of SDGs is produced, which makes it possible to compare the development of different countries. In July 2018, SDSN and Bertelsmann Foundation jointly released *SDG Index and Dashboards Report 2018—Global Responsibilities Implementing the Goals*. It presents a revised and updated assessment of countries' distance to achieving the Sustainable Development Goals (SDGs). It includes detailed SDG Dashboards to help identify implementation priorities for the SDGs. The report also provides a ranking of countries by the aggregate SDG Index of overall performance.

In *2018 SDG Index & Dashboards Report*, 9 goals and 31 index are directly or indirectly related to ecological and environmental protection. *The SDG Index and Dashboards Report 2018—Global Responsibilities: International Spillovers in Achieving the Goals* updated and adjusted the index and methodology on the basis of 2017 data and analyzed the performance of 156 countries in realizing the 17 SDGs, providing reference to the comparison of development level in different countries. In general, among the 17 SDGs and 99 index assessed in 2018, 9 SDGs and 31 index are directly or indirectly linked to ecological and environmental protection (Table 1-2). The 2018 Report added two indicators on the basis of the 29 indicators included in the 2017 Report. To be more specific, in terms of Affordable and Clean Energy (SDG7), access to clean fuels & technology for cooking (% population) and CO_2 emissions from fuel combustion/electricity output (Mt CO_2/TW·h) were added; in terms of Responsible Consumption and Production (SDG12), municipal solid waste [kg/(a·capita)] was added; in terms of Climate Action (SDG13), CO_2 emissions embodied in fossil fuels (kg/capita) was added. Meanwhile, in terms of Clean Water and Sanitation (SDG6), access to quality water (%) was deleted.

Table 1-2   9 SDGs and 31 indicators related to ecology and environment in *The 2018 SDG Index & Dashboards Report*

SDG	Description / Tag
2	Sustainable Nitrogen Management Index
3	Age-standardized death rate attributable to household air pollution and ambient air pollution (per 100,000 population)
6	Freshwater withdrawal as % total renewable water resources
6	Imported groundwater depletion[m^3/(a/capita)]
7	Share of renewable energy in total final energy consumption (%)
7	Access to clean fuels & technology for cooking (% population)
7	CO_2 emission from fuel combustion / electricity output (Mt CO_2/TW·h)
11	Annual mean concentration of particulate matter of less than 2.5 microns of diameter ($PM_{2.5}$) in urban areas (μg/m^3)
11	Improved water source, piped (% urban population with access)

SDG	Description / Tag
12	E-waste generated (kg/capita)
	Municipal Solid Waste [kg/(a/capita)]
	Percentage of anthropogenic wastewater that receives treatment (%)
	Production-based SO_2 emissions (kg/capita)
	Net imported SO_2 emissions (kg/capita)
	Nitrogen production footprint (kg/capita)
	Net imported emissions of reactive nitrogen (kg/capita)
13	Energy-related CO_2 emissions per capita (t CO_2/capita)
	Imported CO_2 emissions, tech-adjusted (t CO_2/capita)
	Climate change vulnerability Monitor (best 0~1 worst)
	CO_2 emissions embodied in fossil fuel exports (kg/capita)
	Effective Carbon Rate from all non-road energy, excluding emissions from biomass (€/t CO_2)
14	Marine sites, mean area protected (%)
	Ocean Health Index-Biodiversity (0~100)
	Ocean Health Index-Clean waters (0~100)
	Ocean Health Index Goal – Fisheries (0~100)
	Fish stock overexploited or collapsed (%)
15	Mean area that is protected in terrestrial sites important to biodiversity (%)
	Mean area that is protected in freshwater sites important to biodiversity (%)
	Red List Index of Species survival (0~1)
	Annual change in forest area (%)
	Imported biodiversity threats (threats per million population)

(3) General performance of SDG Index in countries along the Belt and Road in 2018

2018 SDG Index & Dashboards Report shrank the coverage of countries from 157 to 156. A comparison of the 65 countries along the Belt and Road[①] and the 156 countries shows that except for 4 countries including Brunei, The Republic of Maldives, The Kyrgyz Republic, and the State of Palestine, 61 BRI countries are included in the Report. Table 1-3 lists the index on the fulfillment of 17 SDGs in the

---

① Wang Yiwei. The World is Connected—The Logic of the Belt and Road[M]. Beijing: The commercial Press, 2016:106.

Table 1-3 2018 SDG index scores of 61 countries along the Belt and Road and Their Ranking Among 156 Countries

No.	Country	Score	Ranking	No.	Country	Score	Ranking
1	Slovenia	80	8	32	Turkey	66	79
2	Czech Republic	78.7	13	33	Montenegro	67.6	69
3	Estonia	78.3	16	34	Bosnia and Herzegovina	67.3	71
4	Croatia	76.5	21	35	Tajikistan	67.2	73
5	Belarus	76	23	36	Bahrain	65.9	80
6	The Slovak Republic	75.6	24	37	The Islamic Republic of Iran	65.5	82
7	Hungary	75	26	38	Bhutan	65.4	83
8	Latvia	74.7	27	39	Republic of the Philippines	65	85
9	Moldova	74.5	28	40	Lebanon	64.8	87
10	Poland	73.7	32	41	Sri Lanka	64.6	89
11	Bulgaria	73.1	34	42	Jordan	64.4	91
12	Lithuania	72.9	36	43	Sultanate of Oman	63.9	94
13	Ukraine	72.3	39	44	Mongolia	63.9	95
14	Serbia	72.1	40	45	The Arab Republic of Egypt	63.5	97
15	Israel	71.8	41	46	Saudi Arabia	62.9	98
16	Singapore	71.3	43	47	Indonesia	62.8	99
17	Romania	71.2	44	48	Nepal	62.8	102
18	Azerbaijan	70.8	45	49	Kuwait	61.1	105
19	Georgia	70.7	47	50	Qatar	60.8	106
20	Cyprus	70.4	50	51	Lao PDR	60.6	108
21	Uzbekistan	70.3	52	52	Cambodia	60.4	109
22	China	70.1	54	53	Turkmenistan	59.5	110
23	Malaysia	70	55	54	Bangladesh	59.3	111
24	Vietnam	69.7	57	55	India	59.1	112
25	Armenia	69.3	58	56	Myanmar	59	113
26	Thailand	69.2	59	57	The Syrian Arab Republic	55	124
27	United Arab Emirates	69.2	60	58	Pakistan	54.9	126
28	Macedonia	69	61	59	Iraq	53.7	127
29	Russian Federation	68.9	63	60	Afghanistan	46.2	151
30	Albania	68.9	62	61	The Republic of Yemen	45.7	152
31	Kazakhstan	68.1	65				

61 countries and their global rankings. It could be found that the performance of SDG index differs in Southeast Asian, South Asian, Central Asian, West Asian, Central and Eastern European, Easter European and North African countries. A close look at the 2018 Report reviews a clear difference in the performance of countries in Southeast Asia, South Asia, Central Asia, West Asia, Central and Eastern Europe, Eastern Europe and North Africa.

Central and Eastern European and Eastern European countries rank higher in terms of SDGs Index. 61 countries along the Belt and Road were covered by SDG Index and Dashboards Report 2018, including 16 Central and Eastern European countries and 4 Eastern European countries. In general, 15 of these 20 countries, except for Macedonia, Albania, Montenegro, Bosnia and Herzegovina and Russian Federation, ranked among the top 15 among the 61 countries, outperforming countries in Southeast Asia, South Asia, Central Asia and West Asia. Slovenia (8), Czech Republic (32), Estonia (33), Croatia (34) and Belarus (35) had high rankings among all 156 countries, indicating that they were closer to realizing the 2030 SDGs.

Southeast Asian and Central Asian nations are diverse in their rankings. First, among the 9 Southeast Asian countries, Singapore (43), Malaysia (55), Vietnam (57) and Thailand (59) significantly outperform the Republic of the Philippines (85), Indonesia (99), Lao PDR (108), Cambodia (109) and Myanmar (113). Second, among the 4 Central Asian countries, Uzbekistan (52) and Kazakhstan (65) outperform Tajikistan (73) and Turkmenistan (110).

West Asian nations are sparsely-ranked. West Asian countries have divergent rankings. Among the 19 West Asian nations, Israel (41), Azerbaijan (45), Georgia (47) and Cyprus (50) ranked among the top 1/3 of the 156 countries; Armenia (58), United Arab Emirates (60); Turkey (79), Bahrain (80), Iran (82), Lebanon (87), Jordan (91), Sultanate of Oman (94) and Saudi Arabia (98) ranked among the middle 1/3 of the 156 countries; while Kuwait (105), Qatar (106), Syria (124), Iraq (127), Afghanistan (151) and Yemen (152) ranked among the bottom 1/3 of the 156 countries.

South Asian countries ranked among the bottom 50% of 156 countries in terms of SDG Index rankings. All of the 6 South Asian nations, including Bhutan (83), Sri Lanka (89), Nepal (102), Bangladesh (111), India (112) and Pakistan (126), ranked among the bottom 50%, indicating huge challenges for realizing 2030 SDGs.

1.2.3 Progress on Building a Green Belt and Road

Being green is an important part of the BRI. *Vision and Actions on Jointly Building Silk Road Economic Belt and 21st-Century Maritime Silk Road* states that "we should promote ecological progress in conducting investment and trade, increase cooperation in conserving eco-environment, protecting biodiversity, and tackling climate change, and join hands to make the Silk Road an environment-friendly one". Chinese President Xi Jinping, in his speech at the Legislative Chamber of the Supreme Assembly of Uzbekistan in June 2016, called for the pursuit of green development and efforts to jointly build a Green Silk Road. At the 1st Belt and Road Forum for International Cooperation in May 2017, President Xi Jinping said "we should pursue the new vision of green development and a way of life and work that is green, low-carbon, circular and sustainable. Efforts should be made to strengthen cooperation in ecological and environmental protection and promote ecological civilization so as to realize the goals set by the *2030 Agenda for Sustainable Development*".

In April 2019, President Xi Jinping highlighted at the Opening Ceremony of the 2nd BRF that "The Belt and Road aims to promote green development. We may launch green infrastructure projects, make green investment and provide green financing to protect the Earth which we all call home". In addition, he pointed out that "China and its partners have set up the Belt and Road Sustainable Cities Alliance and the BRI International Green Development Coalition, formulated the *Green Investment Principles for the Belt and Road Development*, and launched the Declaration on Accelerating the Sustainable Development Goals for Children through Shared

1 Green Belt and Road and 2030 Agenda for Sustainable Development

Development. We have set up the BRI Environmental Big Data Platform. We will continue to implement the Green Silk Road Envoys Program and work with relevant countries to jointly implement the Belt and Road South-South Cooperation Initiative on Climate Change".

The essence of green Belt and Road is to integrate green development and ecological and environmental protection into every aspect of the development of the Belt and Road with the principle of energy conservation and environmental protection under the guidance of ecological civilization and green development concepts. First of all, it could be a trigger to promote policy communication with BRI countries; second, it could prevent and control ecological and environmental risks to ensure facilities connectivity with BRI countries; third, it could make industrial capacity cooperation greener to promote unimpeded trade with BRI countries; fourth, it could improve investment and financing mechanisms to serve financial integration with BRI countries; fifth, it could strengthen international cooperation and exchange on environmental protection to promote people-to-people exchange with BRI countries; all the above to provide direct contributions to the realization of environment-related SDGs in BRI countries[①].

As a major initiative driving economic development in related countries, the Belt and Road Initiative has been widely recognized by the international community as an important solution to the implementation of the 2030 Agenda. President of the UN General Assembly Miroslav Lajčák said that China is sharing wealth and best practice through the Belt and Road Initiative to promote the implementation of Sustainable Development Goals. UN Secretary-General António Guterres pointed out that the 2030 Sustainable Development Agenda and the Belt and Road Initiative have the same ambitious goals. They all aim at creating opportunities, bringing beneficial global public products and promote global links in multiple areas including

---

① Dong Zhanfeng, Ge Chazhong, Wang Jinnan, Yan Xiaodong, Cheng Cuiyun. Strategic Implementation Framework for BRI Green Development[J]. Chinese Journal of Environmental Management, 2016,8(2): 31-35, 41.

infrastructure construction, trade, finance, policy and cultural exchange with new markets and opportunities[①]. The Belt and Road Initiative plays an important role in promoting the implementation of the Belt and Road Initiative; for instance, Fred Krupp, President of Environmental Defense Fund (EDF) agrees that EDF could bring economic prosperity and environmental improvement.

The overall objectives and specific tasks and measures are further clarified. In April 2017, The Ministry of Enviromental Protection (now the Ministry of Ecological Environment), the Ministry of Foreign Affairs, the National Development and Reform Commission, and the Ministry of Commerce jointly issued issued *Guidance on Promoting Green Belt and Road*. In May, The Ministry of Enviromental Protection (now the Ministry of Ecological Environment) introduced the *Belt and Road Ecological and Environmental Cooperation Plan*. The *Guidance* pointed out that China will try to establish a practical and highly-efficient system for ecological and Environmental Protection cooperation, support and serve platforms and industrial technological cooperation bases and implement a series of policies and measures on ecological and environmental risk prevention in 3 to 5 years; establish a well- developed ecological and environmental protection service, support and guarantee system in 5 to 10 years. *The Plan* clarified that China would incorporate green development into major activities of the development of Belt and Road, including policy coordination, facilities connectivity, unimpeded trade, financial integration and people-to-people exchange with 25 key projects being listed[②].

A platform for international partnership for green development on the Belt and Road is in the process of establishment. In order to enable BRI countries to better understand the green Belt and Road, China, international organizations, some BRI participating countries and non-governmental organizations have been

---

① UN Secretary-General António Guterres: Belt and Road Initiative has provided new opportunities for global challenges[N]. People's Daily Online, 2017-05-12.
② Cheng Cuiyun, Weng Zhixiong, Ge Chazhong, et al.. Key Tasks for Green Silk Road—*Interpretation of the Belt and Road Ecological and Environmental Cooperation Plan*[J]. Environmental Protection, 2017, 45(18): 53-56.

proactively engaged in seminars, seeking exchanges and coordination on issues related to the green Belt and Road. The Ministry of Ecology and Environment of China and Partners home and abroad jointly launched the BRI International Green Development Coalition, with the goal of improving the capacity of BRI countries on environment governance by building an international platform to exchange ideas, policies and practice and organizing workshops and dialogues, for the purpose to improve the environmental governance along the BRI region. By organizing China-Arab States Environmental Cooperation Forum, China-ASEAN Environmental Cooperation Forum, and China Week for SCO Cooperation, China is proactively engaged in policy dialogues with BRI participating countries.

BRI International Green Development Coalition

In May 2017 President Xi Jinping of China proposed to establish the BRI International Green Development Coalition (BRIGC) in the opening address to the First Belt and Road Forum for International Cooperation (BRF). Under the joint efforts from Partners at home and abroad, BRIGC is officially launched at the Thematic Forum of Green Silk Road of the 2nd Belt and Road Forum for International Cooperation in April 2019. It is an open, inclusive and voluntary international network which will integrate green development into the process of constructing the Belt and Road. It aims to promote international consensus and collective actions of Belt and Road participating countries to implement the *2030 Agenda for Sustainable Development*. By 2019, over 130 Partners from China and the international community joined the Coalition, including 26 environmental authorities from BRI countries, international organizations, research institutions and business, accounting for over 70 international Partners.

Green BRI and 2030 Agenda for Sustainable Development
—Accelerating Effective Alignment and Synergy between BRI and 2030 Agenda to Promote Global Ecological Civilization

The mandates of the Coalition include:

A platform for policy dialogue and communication to:

- share green development concepts and environmental policies;
- provide communication opportunities amongst different stakeholders, and establish a joint research network.

A knowledge and information platform to:

- build an environmental information sharing mechanism;
- provide environmental data and analysis related to the green development of the Silk Road;
- promote capacity building on environment management.

A Platform for green technology exchange and transfer to:

- promote the exchange and transfer of advanced green and low-carbon technology;
- promote investment in green infrastructure and trade.

The Coalition's work will be delivered through a number of Thematic Partnerships made up of coalition partners. The areas of Thematic Partnerships may include, but are not limited to:

- Biodiversity and ecosystem management;
- Green energy and energy efficiency;
- Green finance and investment;
- Improvement of environmental quality and green cities;
- South-South environmental cooperation and SDGs capacity building;

- Green technology innovation and Corporation Social Responsibility;
- Sustainable transportation;
- Global climate change governance and green transformation;
- Environmental laws, regulations and standards;
- Maritime community with a shared future and marine environment governance.

By far, 10 Thematic Partnerships have been initiated. In addition, under the Coalition research on green Belt and Road, a series of seminars and workshops, capacity building activities and pilot projects will be carried out.

The importance of enhancing the corporate environmental and social responsibility of Chinese enterprises operating overseas is emphasized. It has been the focus of UNEP, OECD and World Bank to appeal international investors to follow the high standards of environmental and social responsibility. The UN Global Compact, launched in July 2000, was envisaged to accelerate responsible business action and to ensure that business action and strategy implemented worldwide comply with the ten principles of Global Compact including environment and labor standards, under which the businesses should support a precautionary approach to environmental challenges, undertake initiatives to promote greater environmental responsibility and encourage the development and diffusion of environmentally friendly technologies. Since the 1976, OECD has introduced *Guidelines for Multinational Enterprises* which has been revised several times to underline the dimension of sustainable development by asking enterprises to seriously take the potential environmental impacts of their operation and to strengthen environmental management systems. The World Bank has also adopted environmental safeguards for their financing programs and requires projects to prepare EIAs that meet World Bank standards.

In 2013, Ministry of Commerce and Ministry of Environmental Protection of the People's Republic of China(now the Ministry of Ecological Environment) jointly issued the Guidelines for Environmental Protection in Foreign Investment and Cooperation to guide enterprises to reinforce environmental awareness, perform environmental responsibilities, observe environmental laws and regulations of the host country, conduct environmental impact assessment, implement emergency management and ensure that pollutants' emission meet international standards. In December 2016, 19 global companies of China in the fields of energy, transportation, manufacturing and environment jointly launched an Initiative on Corporate *Environmental Responsibility Fulfillment for Building the Green Belt and Road*. The Chinese enterprises that join the initiative declare that they will observe environmental laws, reinforce environmental management and contribute to green Silk Road in overseas investment and international production capacity cooperation. In April 2019, major financial institutions of China, the UK, France, Singapore, Pakistan, the UAE, Hong Kong SAR and other countries and regions signed up to the *Green Investment Principles for Belt and Road Development*, which marked a new chapter for greening the Belt and Road investment.

Regional environmental governance capacity improves constantly. In December 2015, Forum on China-Africa Cooperation (FOCAC) proposed to implement China-Africa Green Development Plan, to support the implementation of 100 clean energy and wildlife protection projects, environmentally friendly agriculture projects and smart city construction projects in Africa. China-Africa Environmental Cooperation Center will be established to boost the capacity to realize green, low carbon and sustainable development in Africa. In January 2018, China and the Lancang-Mekong countries jointly issued the *Five-Year Action Plan for Lancang- Mekong Cooperation* (2018 — 2022), which calls to establish the Lancang-Mekong Environmental Cooperation Center and implement the Green Lancang-Mekong, focusing on promoting cooperation in air and water pollution control and ecosystem management. In July 2018, the Ministry of Ecology and Environment of China and the Ministry of Environment of Cambodia signed *a Memorandum of Understanding in Phnom Penh*, to jointly establish the Interim Office of China-Cambodia Environmental Cooperation Center, to promote the implementation of project demonstrations in related fields such as domestic sewage treatment and biodiversity protection. In order to push forward environmental capacity building and personnel exchanges among BRI participating countries, the Chinese government embarked on Green Silk Road Envoys Program and South-South cooperation training programs to address climate change for environmental officials, youths, students, volunteers of NGOs, scholars and experts from BRI participating countries. These programs provided support to over 500 delegates from BRI countries to China for discussing issues ranging from environmental impact assessment, air pollution control to water pollution control.

The concept of green economy is being incorporated into BRI. At present, it is the global consensus to develop green economy. UN Environment has also initiated activities on green economy in Africa, Asia-Pacific, as well as Caribbean and Latin America, including research on policy, strategy and indicator system for green economy. In addition, UN Environment is propelling actively the integration of green economy concept into green Silk Road. The Industrial and Commercial Bank of China issued the first Belt and Road Bankers Roundtable Mechanism (BRBR) green bond, and jointly released the Belt and Road Green Finance Index with relevant members of the BRBR mechanism including the European Bank for Reconstruction and Development, the Credit Agricole Corporate and Investment Bank and the Mizuho Bank, to further enhance Belt and Road cooperation on green finance. The China Everbright Group will co-launch the Belt and Road Initiative Green Investment Fund with financial institutions of relevant countries, to solve the problems of insufficient green equity investment and lack of cooperation mechanisms in the BRI participating countries, and to promote the practical innovation of green finance in the BRI participating countries.

Enterprises and environmental NGOs are popularizing high efficiency clean technologies in BRI countries. China has been vigorously promoting the use of high efficiency clean technologies in BRI projects. In September 2016, *the Belt and Road Science, Technology and Innovation Cooperation Action Plan,* jointly issued by Ministry of Science and Technology, National Development and Reform Commission, Ministry of Foreign Affairs, Ministry of Commerce, came into force. The Plan sets out that energy efficiency and emission reduction should be fully integrated into the key areas for technological cooperation, including joint development and demonstration of agricultural technologies, equipment and machinery such as energy and water-efficient agriculture, dissemination of environmental friendly and climate-smart agricultural development model, to promote cooperation in renewable energy such as solar energy, biomass energy, wind energy, ocean energy, and hydropower suitable for countries along the route, and promotion and demonstration of R&D on modern and efficient use of conventional energy such as coal, oil and gas, promotion of cooperative development of new energy vehicle and sharing of data, technology and experience in coping with extreme weather, geological disasters, flood and drought. China and the United Nations Development Programme jointly carried out the BRI Sustainable Investment Facility Project and conducted pilot projects in such countries as Ethiopia. Ministry of Ecology and Environment and the Shenzhen Municipal Government jointly built the Belt and Road Environmental Technology Exchange and Transfer Center (Shenzhen), set up Demonstrative Bases for International Cooperation in Environmental Industries in Yixing, Jiangsu and Wuzhou, Guangxi, forging the platform for Cooperation on Environmental Industries and Technologies between China and BRI participating countries.

Steady progress is achieved in marine environmental cooperation. China has set up marine cooperation mechanisms with Thailand, Malaysia, Cambodia, India and Pakistan. At present, the construction of Thailand-China Joint Laboratory for Climate and Marine Ecosystem, China-Pakistan Joint Marine Research Center and China-Malaysia Joint Marine Research Center are well underway, which focus on cooperation in marine and climate change observation research, marine and coastal line protection, marine resource development and utilization, typical marine ecosystem protection and restoration, and endangered marine species protection.

Green financing safeguards the development of green Silk Road. In order to support BRI, the Chinese government pledged USD 40 billion for the creation of the Silk Road Fund in the end of 2014. In May 2017, at the opening ceremony of the Belt

1 Green Belt and Road and 2030 Agenda for Sustainable Development

and Road Forum for International Cooperation, China pledged to contribute an additional RMB 100 billion to the Silk Road Fund to scale up the support for Belt and Road development. By November 2018, the Silk Road Fund has signed 34 projects with a commitment of approximately 12.3 billion U.S. dollars. Silk Road Fund advocates green, environment-friendly and sustainable development and supports green financing and green investment.

1.3 Promoting the green development of the Belt and Road Initiative, transforming opportunities into reality

Promoting the green development of the Belt and Road requires the establishment of an integrated decision-making mechanism for environmental protection and development in the construction of the Belt and Road. Green development and ecological and environmental protection needs to be integrated into every aspect of the development of the Belt and Road. Policies and operable guidelines on greening the Belt and Road in line with the principles of international cooperation and the implementation of the 2030 SDGs need to be developed to directly help countries along the Belt and Road to realize SDGs related to environmental protection and social development.

Greening BRI requires understanding the impacts that current and planned BRI projects will have on the local and global environment, as well as on sustainable development. It is fundamental to have a specific analysis of the local and global impacts of BRI projects on the environment and sustainable development; this will help to identify which categories of projects should get political and financial attention in upcoming years. Showing the impacts that companies and investors are having through their involvement in BRI projects, both positive and negative, can create significant opportunities to change behaviours and ensure that and ensure that environmental-friendly and sustainable choices are made [①].

[①] China Dialogue.net. A green BRI is a Global Prospective[EB/OL]. [2017-12-15]. https://chinadialogue.net/en/business/10299-a-green-bri-is-a-global-imperative/.

Greening BRI requires applying the principle of Green Infrastructure: The United Nations *2030 Agenda for Sustainable Development* has identified the construction of sustainable infrastructure as one of the 17 major development goals. Investment and construction of green and sustainable infrastructure are the needs of the BRI countries to cope with global challenges such as climate change, resource shortage, and population growth. The BRI should build infrastructure that can promote low-carbon and environmentally sustainable development, such as renewable energy power plants and public transportation systems, and fully consider environmental, social, and governance (ESG) elements during project planning, construction, and operation. It should also maximize the positive role of infrastructure in achieving sustainable development goals, improving economic benefits, taking into account environmental and social considerations, and enhancing the ability to resist natural disasters.

The promotion of green development in BRI projects and all involved countries requires financing, at early stage, greener projects. Green finance has been defined as fundamental support for environmental improvement, climate change mitigation and adaptation, resource conservation and efficient use of economic activities; namely financial services for project investment and financing, project operation and risk management in such fields as environmental protection, clean energy, green transportation, green building and sustainable agriculture. Constantly implementing green financing could direct the flow of capital to greener industries and projects.

Greening BRI requires greening the industrial chain and value chain to build a green supply chain system. Promotion of green production, green procurement and green consumption to drive upper-stream and lower-stream industries to take energy conservation and environmental protection measures and reduce ecological and environmental impact is needed.

Greening BRI requires cooperation in green, efficient and environmental protection technologies and industrial processes to provide effective plans for

environmental governance. It is necessary to enhance the sharing of best practice in the application of environmental protection technologies to promote capacity building in pollution prevention and treatment, facilitate the transfer and development of environmental protection technologies and help countries along the Belt and Road to develop clean industries according to their specific needs and conditions and promote effective pollution control technologies.

1.4 The potential contribution of Green Belt and Road to the implementation of the 2030 SDGs

In 2015, the adoption of the *2030 Agenda* for Sustainable Development and *Paris Agreement* offers a roadmap for a new era of sustainable development. China has committed to building a Green Silk Road. Building a Silk Road corresponds to the international green trend and is consistent with the *2030 Agenda* as well as *Paris Agreement*. Through integrating green development and eco-environmental protection into every aspect and the whole process of the development of BRI, greening the Belt and Road directly contributes to the realization of the 2030 SDGs in BRI participating countries and brings these countries with crucial opportunities for sustainable development.

Promoting ecological and environmental protection policy coordination, strengthening sustainable development partnership. Policy coordination is the foundation of the development of the Belt and Road Initiative. Integrating the development of Green Belt and Road into the framework of SDGs could effectively address the needs and concerns of different countries, realize policy alignment with SDGs and improve the consistency of regional policies on sustainable development. In this way, BRI participating countries could strengthen global sustainable development partnership and integrate projects related to sustainable development into national and local agendas for sustainable development according to their own needs.

Greening the Silk Road under the framework of SDGs could help BRI participating countries to drive economic growth with green projects, avoid stepping onto the old path of "pollution before treatment", and properly handle the relation between economic development and eco-environmental protection. As an advocator of green development, China has the ability to form environmental consensus among BRI participating countries and offer its own solution to realizing environment-related SDGs in the region. Ecological civilization construction could help to improve the capacity of governments in the transition to a green development model, while a green Silk Road enables China to make joint efforts with BRI participating countries to promote ecological civilization construction worldwide. In this way, developing countries along the Belt and Road could integrate SDGs into their national and local development plans and give due consideration to environmental, social and economic factors in designing projects.

China will strengthen the construction of ecological and environmental protection cooperation mechanisms and platforms to carry out high-level intergovernmental dialogues with BRI participating countries and use cooperation mechanisms including China-ASEAN, Shanghai Cooperation Organization, Lancang-Mekong, Euro-Asia Economic Forum, Forum on China-Africa Cooperation and China-Arab States Cooperation Forum to strengthen regional communication and exchange on ecological and environmental protection. Policy coordination and exchange could effectively promote the development of sustainable partnerships among BRI countries (SDG Target 17.16) and improve the consistency of regional policy on sustainable development (SDG Target 17.14).

Reducing the environmental risks brought by facilities connectivity, protecting regional ecological systems. Green Silk Road requires green infrastructure construction, which means constantly promoting ecological and environmental friendly public goods and infrastructure construction for environmental protection, promoting the development and transfer of environmental protection technologies, exchange and cooperation between clean industrial parks and ensure the adoption of clean technologies in infrastructure construction. Green facilities connectivity is linked to multiple SDGs and targets. It could protect and sustainably use terrestrial ecosystems and their services (SDG Target 15.1), reduce the degradation of natural habitats and halt biodiversity loss (SDG Target 15.5) and help related countries to upgrade infrastructure with increased resource-use efficiency and greater adoption of clean and environmentally sound technologies and industrial processes (SDG Target 9.4).

Sustainable infrastructure construction in BRI participating countries could help to develop an integrated system for ecological and environmental risk prevention and governance, promote joint efforts in developing incentive mechanisms for investment in sustainable infrastructure along the Belt and Road, establish an open database for sustainable infrastructure projects, and promote the implementation of best practices in environmental and infrastructure planning.

Promoting green unimpeded trade, improving the efficiency of production and consumption. In trade, green Belt and Road will facilitate environmental-friendly product and service trade, make the market of environmental services more open and expand the import and export of environmental products and services. International cooperation on green supply chains is an important measure, which promotes green development throughout the industrial chain from production to product flow and to consumption through the development of Belt and Road green supply chain cooperation platforms. These activities could increase sustainable production and consumption in countries along the Belt and Road through trade and help BRI countries to gradually improve resource-use efficiency in global consumption and production (SDG Target 8.4).

Promoting green financial integration, encouraging investment in clean technologies. Green Belt and Road needs financial tools to identify and prevent social and environmental risks brought by related projects, improve environmental information disclosure, strengthen project environmental risk management and make foreign investment greener. Green investment in countries along the Belt and Road has attracted wide attention with promising prospect for the development of green industries[①]. Green finance could effectively promote investment in energy infrastructure and clean energy technologies (SDG Target 7.A)[②] and mobilize additional financial resources for developing countries from multiple sources (SDG Target 17.3). For example, the Silk Road Fund has been implementing the concept of green development and green finance with emphasizing green, environmental-friendly and sustainable development as one of the investment principles and promoting the interaction of clean and renewable energy on multiple levels for extensive cooperation in energy and resource as one of the four investment priorities.

Take clean energy as an example, China has become the world's largest producer of clean energy equipment and market of renewable energy. Clean energy businesses and technologies from China are flourishing in BRI participating countries, becoming a major driver for global energy transition (Figure 1-1). According to Brookings Institution, as a game changer of the global clean energy industry, China will play a crucial role in promoting a transition to non-fossil fuels in global energy consumption.

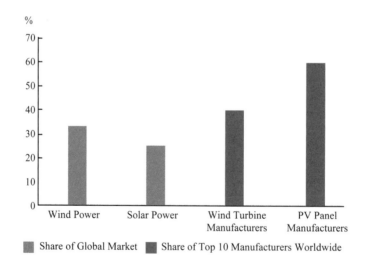

Figure 1-1 The share of China in the global clean energy market

Source: International Energy Agency.

---

① Belt and Road: Promoting new Opportunities in the Development of Green Finance[R/OL]. [2018-07-09]. http://greenfinance.xinhua08.com/a/20180709/1768414.shtml.
② Goal 7.A: By 2030, enhance international cooperation to facilitate access to clean energy research and technology, including renewable energy, energy efficiency and advanced and cleaner fossil-fuel technology, and promote investment in energy infrastructure and clean energy technology.

Still having huge demand for coal and oil, China plans to invest more than 6 trillion USD in low-carbon power generation and other clean energy technologies. So far, China has issued nearly 25 billion USD of green bonds in infrastructure construction, covering areas such as clean energy, clean transportation, pollution prevention and control, and energy conservation technologies.

Strengthening people-to-people exchange in environmental protection and, promoting capacity building in developing countries. Green Belt and Road will strengthen the support to green demonstration projects, promote exchange and cooperation in environmental protection policy, legal system, talent training and demonstration projects, continue to carry out the Green Silk Road Envoy Plan, increase the interaction and communication between environmental management experts and technical experts in BRI countries, promote environmental protection technology and industrial cooperation and improve the environmental protection capability of BRI countries. These activities and projects will effectively support developing countries to improve technological and technical capabilities, adopt more sustainable production and consumption patterns (SDG Target 12.A), promote the development, transfer, dissemination and diffusion of environmentally sound technologies to developing countries (SDG Target 17.7) and enhance international support for implementing effective and targeted capacity-building in developing countries through South-South cooperation (SDG Target 17.9).

2 Opportunities and Challenges in the Development of Green Belt and Road

2.1 Opportunities

China, a prominent green development player domestically, has the capacity to promote environmental convergence among the BRI participating countries (for example, on circular economy) and provide a solution to realizing environmental-related SDGs in the region. Integrating the important concepts and practical results of China's ecological civilization construction into the BRI not only enriches the connotation of the BRI, but also provides an inevitable choice for the BRI to move towards a high-quality development stage. Building a green BRI can provide a model of green transformation for the participating countries, and promote high-quality economic development through ecological environmental protection.

Green Silk Road will create tremendous opportunities for green development and capacity building in BRI countries, including:

- Promote public awareness of environmental protection and reduce resource consumption, boost green industries and low-carbon lifestyles;
- Help BRI participating countries integrate the SDGs at country, regional and project levels;
- Promote high standards in all BRI projects;
- Construct an integrated risk governance system for the BRI while enabling sustainable development;
- Engage with policy-makers in BRI participating countries to establish frameworks that incentivise sustainable BRI infrastructure investments that are currently not financially viable and set up an open access database for sustainable BRI infrastructure projects to implement best practice in environmental and infrastructure planning;

2 Opportunities and Challenges in the Development of Green Belt and Road

- Set up a cross-sector "Greening the Belt and Road" learning and leadership platform to draw attention to the environmental risks and opportunities and ways to respond to them;
- Support efforts to create an open and optimized policy environment and constantly improve transparency.

2.2 Challenges

Since the 21st Century, sustainable development has become the theme of the international community with important progress achieved in the field of international environment and development. Meanwhile, the environmental pressure brought by human beings on earth continued increasing. Among the three pillars that support the realization of sustainable development goals, performance on the environmental dimension still lagged behind the social and economic progress. There are even regresses in the international environment and development field.

The Belt and Road Initiative is the largest infrastructure program ever planned. It comes with both opportunities and risks – for investors, for sustainable development, and for natural resources (Figure 2-1). At the same time, although sustainable development has become a global consensus, the inability of BRI participating countries in promoting ecological and environmental protection and the complexity of international cooperation projects bring a series of challenges to greening the Silk Road.

Sector	Infrastructure	Inputs			Outputs				
		Ecosystem use	Water use	Other resource use	Greenhouse gas (GHG) emissions	Non-GHG air pollutants	Water pollutants	Solid waste	Total Impact
Energy	Coal power plants								
	Hydro plants								
	Gas-fired power plants								
	Pipelines								
	Solar power plants								

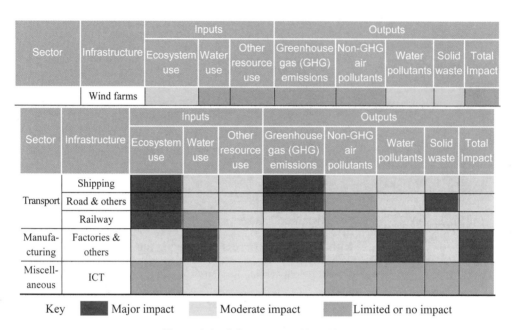

Figure 2-1　Infrastructure Heat Map

Source: Greening the Belt and Road Initiative report, HSBC-WWF, 2017.

In terms of green-oriented philosophy, in many BRI participating countries, the introduction of low carbon technologies & environmental design requirements are disregarded or best in early-stage. Many countries are under-developed and they never had the chance before to get steadily in touch with such concepts. Moreover, they have the understandable desire to develop fast, talking "green issues" in a second stage.

In terms of policy and monitoring assessment, BRI projects are mostly very complex and transnational; they involve different standards and procedures to plan, design, construct, operate and assess projects. Moreover, the business scenario for investing in green or natural infrastructure is often not clear. In some cases local laws and technical standards are very vague or completely missing. The number and variety of sustainability standards and assessment methods makes it difficult for financial investors to ensure they invest only in sustainable infrastructure. Returns are often not ideal, and affect the enthusiasm of investors.

In terms of information and transparency, in particular transnational BRI projects tend to be extremely difficult; data is scattered and difficult to locate as planning (in some cases completely missing), designing and implementing BRI projects is mostly decentralised. The complexity of BRI projects requires precise planning and complete transparency in its execution.

In terms of green project's implementation, so far, most of the developed projects cannot be defined as "green" at least according to international standards; moreover, as most of involved BRI countries are under-developed, it is mostly requested to quickly develop projects so to ensure fast economic and social development without taking too much into consideration negative environmental impacts of projects developed within BRI. International and private companies feel too exposed in getting directly involved in BRI projects.

In terms of green finance and green investments, so far they have not yet gained adequate attention among finance and wider private sector players or broader stakeholders. There is not yet a general acceptance in the definition of green projects and consequently in defining what can be financed within present green investments, moreover ROR and ROI in Green projects is limited connected to the expected risks in the long run.

3 Major Issues on the Belt and Road

3.1 Long-term mechanism of constructing green Belt and Road

The long-term mechanism of constructing green Belt and Road includes strategy connection, embedding green Belt and Road into the overall planning and the project review mechanism among other aspects.

3.1.1 Research on the strategy connection mechanism

(1) The current state of strategy connection on building a green Belt and Road

The strategy connection mechanism marks an important foundation for building a green Belt and Road. In the past 6 years, centering on themes of green Belt and Road and sustainable development, China and BRI Participating Countries hosted a series of bilateral and multilateral dialogue and exchange activities focusing on various fields and at multiple levels, which attracted the active participation of government officials, think tank experts, media representatives and entrepreneurs from China and BRI participating countries as well as representatives of relevant international organizations. These activities include the Ecology Session of Euro-Asia Economic Forum, China-Arab States Forum on Environmental Protection Cooperation, and China-ASEAN Environmental Cooperation Forum, etc.

Since 2015 China has signed nearly 50 bilateral and multilateral ecological and environmental cooperation documents with BRI participating countries. According to these documents, both parties will cooperate in the fields of environmental policy dialogue and information sharing, pollution prevention and control and environmental governance, ecosystem management and biodiversity protection, environmental industry and technology, environmental monitoring, environmental education and

3 Major Issues on the Belt and Road

public participation, and other areas related to the protection and improvement of the environment as agreed by both parties in line with their cooperation needs. The cooperation and exchange will be implemented with sharing information and materials, exchanging experts, scholars, and delegations, and organizing themed seminars, workshops, or joint meetings.

(2) Problems of the strategy connection mechanism on building a green Belt and Road

Firstly, the knowledge and awareness gap in the importance of ecological and environmental issues increased the difficulty of making connections. As BRI Participating countries are at different development stages with various degrees of knowledge or resource input on ecological conservation and environmental protection, the environmental governance regulations and policies of different countries vary. For some countries, their related laws, regulations, and technical standards are not complete, which made it more difficult to realize strategy connection since participating countries cannot adopt comprehensive strategies or standards when carrying out strategy connection on green Belt and Road.

Secondly, there is no special mechanism under the green Belt and Road framework established to promote strategy connection. The current strategy connection is realized mainly based on the Belt and Road Forum for International Cooperation and Environment Ministers' Meeting under China-ASEAN, SCO, and BRICS cooperation mechanisms. Although ecological and environmental departments are actively facilitating BRI participating countries to sign MOUs on environmental cooperation, there has not form a special mechanism under the green Belt and Road framework.

3.1.2 Research on the mechanism for embedding green Belt and Road into the overall planning

(1) Overview of green Belt and Road planning

As a significant special planning under the 13th Five-Year Plan of China's national economic and social development, *China's 13th Five-Year Plan for Ecological and Environmental Protection* raised clear requirements on green development on the Belt and Road, specially put forward contents on promoting green development on the Belt and Road, and made comprehensive arrangements on the efforts of promoting the construction of green Belt and Road during 2016 – 2020.

In addition, in terms of the construction of the "Six Economic Corridors" on the Belt and Road, *the Plan Outline of Constructing China-Mongolia-Russia Economic Corridor* released in June 2016 incorporated tasks on enhancing ecological and environmental cooperation, including studying on the feasibility of establishing information-sharing platforms, expanding cooperation in disaster prevention and mitigation, and carrying technological exchange and cooperation in the domain of ecological conservation and environmental protection. *The China-CEEC Medium-term Cooperation Plan* released in November 2015 also includes cooperation on ecological and environmental protection and sustainable development.

(2) Insufficiencies of the embedding mechanism on green Belt and Road

Firstly, ecological and environmental issues have not become a priority cooperation area in related cooperation planning on constructing the Belt and Road. In the current cooperation planning on the Belt and Road, ecological and environmental protection and green development haven't been listed as priority cooperation contents judging by related text size or statement on its significance. For instance, among the 58 cooperation items in the aforementioned *China-CEEC Medium-term Cooperation Plan*, only the 39th and 40th Item are related to ecological and environmental cooperation. It is a sign that all parties consent that infrastructure connectivity, international capacity cooperation, and construction of economic and trade cooperation zones outside of China are the priority areas of cooperation.

Secondly, overall bilateral and multilateral cooperation plan for ecological and environmental protection under the framework of the Belt and Road has not be established yet. At present. China mainly carried out cooperation on constructing the Belt and Road by signing MOUs or inter-governmental agreements with participating countries. China worked with some countries (such as Kazakhstan) to formulate planning in certain fields (such as international capacity cooperation), but haven't developed a cooperation plan for ecological and environmental protection as well as green development.

3.1.3 Research on the project evaluation mechanism

(1) The current state of Belt and Road project management in China

At present, the cooperation projects on the Belt and Road are still managed according to the original management methods and channels, facilitated by related departments and industry organizations, and specifically implemented by the business. The National Development and Reform Commission and other relevant departments support foreign investment projects with enterprises as the main body, market-oriented, and in accordance with commercial principles and international practices, especially supporting enterprises to invest and operate BRI and international capacity cooperation projects. According to the *Guiding Opinions of the State Council on Promoting Cooperation on International Capacity and Equipment Manufacture* (Guofa〔2015〕No. 30), railways (extending to infrastructure), electricity, automotive, communications, are the key industries for international capacity cooperation.

(2) The ERA problems of the Belt and Road projects

The weights of ecological and environmental influence and risks in project evaluation and decision-making need to be improved. At present, evaluations on the Belt and Road construction projects are mainly measured based on economic feasibility, technological availability, and a leading influence on local economic and social development. Projects' influence on the ecology and environment is only incorporated into evaluation and consideration as risk factors. As there are no unified and regulated evaluation methods and standards on projects for industry sectors and financial agencies, there is still room for improvement in the weights of ecological and environmental influence and risks of projects. The BRIGC launched a research project on Green Development Guidance for BRI Projects at the end of 2019, aiming to formulate project classification and evaluation plan for BRI projects, assess the ecological environment and climate impact, and further identify stakeholders, thus provide guidance on the ecological environment risks of overseas investment.

3.2 Strategic arrangement and implementation mechanism of green finance in the development of the Belt and Road Initiative

Green finance facilitates the development of the Belt and Road Initiative through addressing two issues: environmental risk management and green investment and financing. In terms of environmental risk management, green finance helps financial institutions to integrate environmental protection, ecological conservation and climate change mitigation into the decision-making process; in terms of green investment and financing, green finance helps financial institutions to increase the investment in green, low-carbon and circular economy projects with streamlined procedures. Financial institutions needs to develop green finance or green credit, as has been indicated in the *Green Credit Guidelines* issued by the former China Banking Regulatory Commission (in 2018, it merged with the China Insurance Regulatory Commission to form the China Banking and Insurance Regulatory Commission, referred to as the China Banking and Insurance Regulatory Commission) in February 2012 and *Guidelines for Establishing the Green Financial System* issued by seven ministries, including the People's Bank of China and NDRC, in August 2016. To effectively support the development of the Belt and Road Initiative, Chinese financial institutions not only need to implement the two documents, but also adopt new approaches to developing green finance that are different from those adopted in the domestic market. Chinese financial institutions have two objectives: integrate the Sustainable Development Goals (SDGs) into the development of the BRI through finance; promote the good practice of China in green finance based on the reality and investment environment of countries and regions along the Belt and Road.

3.2.1 Good Practice and Experience of Green Finance Worldwide

Requirements for financial institutions worldwide on promoting environmental risk management come from four aspects:

First, international rules and international cooperation. Many international organizations and regional multilateral institutions have developed or promoted the formation of international standards on environmental and social governance, with much efforts being taken to promote dialogue among countries and institutions. Examples include United Nations Global Compact, Equator Principles and *United Nations Principles for Responsible Investment*.

Second, rules and standards developed by multilateral developmental financial institutions. Based on their tenants or rules, developmental financial institutions have developed and issued their own standards and guiding principles, such as Inter-American Development Bank, World Bank Group, Asian Development Bank and African Development Bank. These standards and instruments have been adopted, to a large extent, by governments and financial institutions around the world.

Third, national policies and regulations. This mechanism could manifest itself in different ways. Typically, environmental laws and regulations would require environmental impact assessment for development projects, identify protected areas and define the upper limit of pollutants. Besides, regulations on the green development of financial institutions have been tightened and improved in recent years through restrictions on investment in certain areas/industries and incentive measures for green investment. Examples include *Green Credit Guidelines* and related policies issued by China, *Policy for Socio-Environmental Responsibility* issued by the Central Bank of Brazil in 2017, *Sustainable Banking Principles* issued by the Central Bank of Nigeria in 2012 and *Guidelines on Environmental & Social Risk Management* (ESRM) issued by the Central Bank of Bangladesh in 2017.

Fourth, voluntary commitment of commercial financial institutions. To pursue sustainable development and keep in line with domestic and international trends, commercial financial institutions would make voluntary commitments to comply to environmental and social standards to improve its environmental and social performance and enhance environmental risk management in investment and operation.

In spite of the varying sources and forms, environmental risk management is manifested as the adoption of sustainable development as the strategic priority by financial institutions, the integration of environmental considerations into the process of investment and decision-making and the disclosure of investment and credit policies for environmental-sensitive industries.

Green investment is often guided by the following four factors.

First, green investment catalogues or standards, which is the major approach adopted by China to encourage green investment.

Second, green investment institutions established and funded by the country committed to guiding the market and commercial institutions to make green investment, UK Green Investment Bank is a typical example.

Third, priority considerations and key areas for investment identified by financial institutions based on their own strategies and tenants. World Bank, for example, highlights in its 2018 Annual Report that the World Bank supports its client countries in three priority areas: promoting sustainable, inclusive economic growth; investing more-and more effectively-in people; and building resilience to fragility, shocks, and threats to the global economy. Commercial financial institutions such as HSBC and Citibank have also developed strategies to increase investment in renewable energy and low-carbon infrastructure.

Fourth, standards initiated and widely recognized by industrial associations and research institutions with green bond as a typical case. The *Green Bond Principles* developed by International Capital Market Association (ICMA) and some standards and tools developed by Climate Bonds Initiative have been gradually accepted and adopted by financial institutions.

By the same token, the substantive manifestation of promoting green investment is a growth in investment in green, low-carbon and circular economy projects through product, institutional and procedural innovation with controlled risks.

3.2.2 The Institutional Arrangement of Environmental Risk Management in the Development of Green Belt and Road

The Chinese government has been attaching great importance to controlling the environmental and social risks of overseas operations of financial institutions[①]. The former China Banking Regulatory Commission (CBRC, later merged with the China Insurance Regulatory Commission to form the China Banking and Insurance Regulatory Commission in 2018. Now it is the China Banking and Insurance Regulatory Commission, or CBIRC) issued *Green Credit Guidelines* in February 2012, promoting banking financial agencies to use green credit as an instrument for taking active measures to adjust the credit structure, effectively prevent environmental and social risks, better serve the real economy, and catalyze the transformation and restructuring of economic development. In June 2014, CBRC released *Key Indicators on Evaluating the Green Credit Implementation* as a move of implementing regulations like *Green Credit Guidelines*. The Indicators set up quantitative and qualitative indicators from the perspective of organization management, policies, and mechanism, capacity building, process management, internal control management, and information disclosure, supervision, and inspection among other aspects to guide banking financial agencies to carry out self-evaluation over implementing green credits. In January 2017, CBRC released *Guiding Opinions on Regulating Banking Industry to Serve Companies Going Global and Enhancing Risk Prevention and Control*. Article 5 prescribes that the financial agencies in the banking industry should continuously improve comprehensive risk management capacities, including environmental and social risk control. In particular, banks and financial institutions are required to pay special attention to environmental and social risks in the fields of energy resources, agriculture, forestry, animal husbandry and fishery, major

---

① Environmental Risk Management Initiative for China's Overseas Investment[R/OL]. [2017-09-06]. http://www.xinhuanet.com/money/ 2017-09/06/c_1121612075.htm.

infrastructure, and engineering contracting when providing project financing and trade financing.

In August 2016, 7 ministries and commissions including the People's Bank of China, the Ministry of Finance, and then the Ministry of Environmental Protection (now the Ministry of Ecological Environment) jointly issued the *Guiding Opinions on the Construction of a Green Financial System* (hereinafter referred to as the Guiding Opinions), which strengthened China's green financing policy foundation from the perspectives of environmental information disclosure and market regulatory rules. The 8th Part of Guiding Opinions proposed to widely carry out international cooperation in the area of green finance and facilitate the improvement of green foreign investment, encourage and support Chinese financial agencies, non-financial companies and multilateral development institutions that China is involved to strengthen green development conceptions during investing and constructing Belt and Road projects. They need to improve the capacities of embedding green and low-carbon development in terms of environmental risk management, environmental information disclosure, green supply chain management, and environmental pollution obligation insurance.

In May 2017, then the Ministry of Environmental Protection (now the Ministry of Ecological Environment) released the *Belt and Road Ecological and Environmental Cooperation Plan*, which regards facilitating green investment and financing as an integral part of building a green Belt and Road. Constructing the green financial system under the Belt and Road framework should start from national strategy deployment, adhere to the conception of sustainable development, encourage and support more financial resources into green industries, use green bonds, credit and other financial tools to control and reduce polluting investment, complete standards, guidance, and process management in the environment, society and governance, supervise more financial agencies to effectively recognize, evaluate and control environmental and social risks in investment and financing activities, and form policy synergy towards a green economy.

Documents mentioned above listed strict and detailed regulations and requirements on "greening" and environmental risk prevention and control. However, the implementation of these rules are far from satisfying. Major problems include:

First, standards for green and environmental-friendly investment have not been fully unified, which increases the investment risks. Currently, most Chinese financial institutions have not joined any international codes. Therefore, the environmental policies being implemented in international financing by Chinese financial institutions could be seen as an extension of domestic policies on green credit. The common practice is to adopt Chinese laws and regulations or refer to related standards of the hosting country. However, the standards designed for the Chinese market may not be applicable in the hosting country. International investment and cooperation have greater uncertainties due to differences in political, social and cultural realities, legal systems, values and investment environment.

Second, the policy framework of green finance need to be improved, or it will influence the effect of implementation. Government departments play a leading role in the development of green finance for the Belt and Road Initiative. They are expected to provide concrete measures for the overseas implementation of green finance. Varied execution standards for different financial agencies and influencing specific business on green investment and financing business, has also influenced the quality of statistics data. Meanwhile, certain technical factors restrained the development of green finance, including insufficient feasibility of environmental data, lack of analysis methods on environmental risks, or incomplete information disclosure platforms. In project approval, relevant departments failed to take environmental considerations as important factors in selecting projects. Besides, related policies having been launched couldn't effectively guide foreign direct investment to environmental-friendly projects as they don't have enough binding force.

Third, the implementation of green financial policies needs to be more carefully aligned with greater operability. As important participants in the development of the

Belt and Road Initiative, Chinese financial institutions still couldn't meet international standards in the environmental and social management of overseas projects. The degree of matching or aligning with international standards or conventions needs to be enhanced. In the process of policy implementation, Chinese financial institutions usually replace compliance assessment with quantitative environmental risk assessment and integrate environmental policies into financing decision-making based on "veto" and "blacklist" without clear, effective and quantitative detailed rules for the credit issuance or defined assessment standards for overseas investment projects.

Fourth, corporate awareness in environmental protection and effective supervision should be further enhanced. In the implementation of green finance, improvement in corporate awareness, the guidance of governments and the participation of banks are equally important. Currently, most Chinese businesses engaged in the development of the Belt and Road Initiative have accumulated relatively rich experience in environmental protection and social responsibility. However, some of them failed to take their environmental responsibility. Even though China has issued some policies, guidance, and suggestions on foreign green financing and investment, restraints imposed on non-green financing and investment behaviors are not strong enough. The environmental monitoring system on foreign investment management has not been formed. There is a lack of environmental laws and regulations on foreign investment, especially special environmental laws and regulations that regulate foreign investment behaviors of companies. The key to improving environmental risk management in BRI investments lies in strengthening legal and administrative supervision and impose stricter punishment for behavior with environmental impact that cause business loss.

Fifthly, there is not enough support from guiding capitals to formulate persuasive attraction to the market. Green investment and financing are involved with complicated and constantly renewed special technologies involving environmental risk assessment and carbon trade, which poses a very high requirement on the risk evaluation and management capacity of financial institutions. To this end, it is necessary to prepare some guiding capitals and fund for capacity building to support relevant government departments with experience to work on special fund management and capacity building, etc., so as to gradually attract market entities to join.

3.2.3 Institutional Arrangement of Green Finance in the Belt and Road Initiative

Currently, China has developed detailed rules and standards for green investment at home, including *Statistical System for Green Credit* issued by CBRC, *Catalogue of Projects Supported by Green Bond* issued by the People's Bank of China, and *Guidelines for the Issuance of Green Bond* issued by NDRC. *Statistical System for Green Credit*, in particular, identified the catalogue of eleven green industry categories and projects and made it clear that "overseas projects adopting international practices or international standards" are also within the scope of Statistical System for Green Credit. Green bond issued by Industrial and Commercial Bank of China, Agricultural Bank of China, Bank of China and Industrial Bank overseas also comply to the rules on the issuance of green bond in China.

Currently, investment in BRI green projects is still in the primary stage with limited successful cases. Four issues need to be addressed for further improvement.

First, the standards of BRI green projects. Three issues needs to be clarified in terms of the standards of BRI green projects. The first is the scope of green projects. Whether they are projects engaged in the development of green, low-carbon and circular economy or projects in key industries or areas needs to be further clarified. The second is the definition of green standards. We need to decide whether to adopt domestic standards on green finance and green credit or the best practice of international institutions, multilateral banks and commercial financial institutions, and whether to use the standards of hosting countries or develop new standards for BRI green projects. The third is the quantitative assessment of green standards. We need to decide whether to identify green projects with the type and nature of the projects or with the effects, impact and contribution of the projects.

Second, risk control for the investment environment in countries along the Belt and Road. Unclear business and political rules, market and exchange rate risks caused by uncertainties are the major challenges hindering the implementation of BRI projects. This is caused by several reasons: first, countries along the Belt and Road have different political, social, religious and cultural realities; second, financial institutions lack the experience and expertise for non-financial risk management; third, the coordination mechanism on business and trade between China and related countries have to be improved.

Third, the business return of BRI green projects. High cost, huge investment, low short-term ROI and long investment cycle contribute to the financing difficulties faced by green projects. In the development of BRI, Chinese financial institutions lack effective information and understanding of overseas projects, which adds to the uncertainty on the return of green projects.

Fourth, the incentive mechanism for the investment in BRI green projects. The fundamental reason for the lack of investment is the lack of incentives. Financial institutions could not enjoy preferential policies of hosting countries in investing BRI green projects, without extra business return. To address the issue, we need to encourage and support the development of BRI green projects with risk compensation, credit guarantee, favorable tax policies and subsidized loans and motivate financial institutions for more effective implementation of green credit.

3.3 The Belt and Road and constructing a more inclusive global green value chain

3.3.1 The significance of building green value chain in greening the Belt and Road

The core of green value chain. In the past 40 years, with the liberalization and facilitation of investment and trade, international division of labor has transformed from interindustry division of labor to intra-industry division of labor. With the transformative development of information and communication technologies bringing down the cost of transboundary communication and coordination, global production lines emerged, which leads to the formation of Global Value Chains, or GVCs, as the value created by the production process is distributed around the world. Developed and developing countries alike engage in the process of value creation, distribution and redistribution, covering design, R&D, production, transportation, consumption and recycling, although they play different roles. Green Global Value Chains (GGVC) incorporates green development concepts into GVCs[①], highlighting the environmental and climate impact of and in each process. It forms a closed loop comprising green design, green production, green products, green marketing, green consumption, green recycling and green materials and reflects the environmental footprint of the transboundary transfer of value creation against the background of the global relocation of production.

The development of BRI creates opportunities for the construction of a more inclusive GGVC. BRI links countries in different stages of development and of different historic and cultural backgrounds closely together through promoting connectivity, building an international platform for integration of products and services. However, BRI participating countries are in different stages of economic growth with different environmental and development priorities. Consequently, it is impossible for a unified environmental rule to meet the demands of all BRI participating countries. To address the issue, it is necessary to develop multi-dimensional, multi-layer environmental standards in relation to the actual needs and ecological and environmental conditions of countries in varying stages of development through "rule-based governance" on the basis of connectivity and mutual trust promoted by BRI.

A more inclusive GGVC is crucial to the continuous development of BRI. BRI promotes productive capacity cooperation based on mutually-beneficial win-win

---

① Zhou Yamin. Building a more inclusive global green value chain[N]. People's Daily Online, 2017-05-17.

cooperation for the sound development and share prosperity of BRI participating countries and regions. As an important part of BRI green development, GGVC promotes global division of labor guided by the principles of green development, which enables all participants to take their fair share of the benefits of the industrial chain while encouraging synergistic efforts to reduce the burden on environment. The result is a win-win situation with economic, social and environmental benefits that safeguards the continual and sustainable development of BRI[①].

3.3.2 The necessity of constructing GVCs

As a global public good, environment is an important spectrum of global governance. In the context of fragmented global production, constructing more inclusive GVCs is crucial for greening the Belt and Road.

Constructing a Global Green Value Chain helps developing countries to reduce resource intensity in industrial production and the reliance on resource-based bulk commodities, which is conducive to the sustainable regional development. In the future, through strengthening South-South Cooperation and implementing BRI, constructing Global Green Value Chain has the potential to be an important means for developing countries to realize their own development goals and fulfill international responsibilities. Second, constructing coordinated, inclusive, green and sustainable GVCs is an urgent task that requires all countries to reconsider international investment, production, trade and cooperation from a strategic perspective and reshape Global Value Chains with green and sustainable development concepts.

---

① CCICED. China's Role in the Global Green Supply Chain[R]. 2016-12.

4 Case Studies on Green Development on the Belt and Road

4.1 Field studies on Pakistan and Sri Lanka

During February 20th and 27th, CCICED SPS visited Pakistan and Sri Lanka for field studies. The goal of the field studies to Pakistan and Sri Lanka was to achieve a greater understanding of the ongoing BRI investment in BRI host countries and to identify the environment impacts and challenges of the BRI projects and how best they could be assessed. In addition, to identify lessons to be learnt from the existing BRI projects and look at new opportunities for policy interventions to make BRI greener.

The field study finds that in general the two countries recognize the environmental protection work of Chinese projects. The governments, research institutions, or social organizations of Pakistan and Sri Lanka recognized the environmental protection work of the Belt and Road projects. They held that these projects could abide by local environmental regulations because the governments of both countries set clear environmental management rules over projects. The other reason is that the Belt and Road projects are jointly constructed based on mutual consultations and local cooperative companies maintain relatively high pursuits and considerations on environmental protection. Meanwhile, some social organizations also held that the projects could do a better job in the area of environmental protection. For instance, the European Union organization in Pakistan expressed hope that the environmental information during the implementation of China-Pakistan Economic Corridor projects could be more transparent.

4.1.1 The ecological and environmental state of Belt and Road projects in Pakistan

(1) Overview of the basic situation

The fragileness of various ecosystems in the northern part of Pakistan is recognized globally. These ecosystems includes temperate deciduous forests, coniferous alpine forests, tundra and grasslands. Mountain ecosystems in these areas also include glaciers, which are the most important source of freshwater for local and downstream communities and the cornerstone of regional biodiversity. These ecosystems are also vital to the livelihoods of communities in high altitude areas. Poverty and extreme dependence on natural resources, as well as extreme weather and climate change-related challenges, including melting glaciers and sudden flooding in glacial lakes, have increased the vulnerability of local communities.

The China-Pakistan Economic Corridor is an important part of the "Belt and Road" initiative. The China-Pakistan Economic Corridor extends from Kashgar in the north to Gwadar Port in Pakistan in the south. It has a total length of 3,000 kilometers and has a core influence on China-Pakistan relations. It is the flagship project among the six economic corridors under the "Belt and Road" and is regarded as a "game changer" in the region. The China-Pakistan Economic Corridor is a regional connectivity framework that strengthens regional integration and trade flows, involving the development of transportation, energy, industry and other forms of infrastructure. Since the China-Pakistan Economic Corridor originated in the northern part of Pakistan, the northern area is called the "gateway of the China-Pakistan Economic Corridor" and has huge potential, especially in hydropower and eco-tourism. These areas are also rich in natural resources, including forests, water, glaciers, biodiversity and mineral resources. There are four key pillars of investment in the China-Pakistan Economic Corridor (CPEC), including Gwadar Port, energy projects, infrastructure construction, and industrial cooperation.

The government branch in charge of environmental governance in Pakistan is the

Ministry of Climate Change. Its core legislation and regulation on environmental governance in the *Environmental Protection Law*, which prescribes to carry out comprehensive environmental impact assessments on projects that might bring damages to the environment, with pollution discharged not allowed to surpass national environmental quality standards. The economic benefits of the construction of the China-Pakistan Economic Corridor are significant, but it is also necessary to pay attention to the various beneficiaries and the impact of the project at the local level, especially on local communities. The launch of the China-Pakistan Economic Corridor will bring certain challenges to community livelihoods and fragile biodiversity and ecosystems.

The environmental authority in Pakistan is the Environmental Protection Agency of the Ministry of Climate Change. The core environmental management regulations are the *Environmental Protection Law*, which stipulates that comprehensive environmental impact assessments should be conducted for projects that may cause environmental damage, and the amount of pollutants must not exceed the national environmental quality standards. CPEC has a relatively complete negotiation mechanism: planning cooperative work or implementing specific projects will be carried out by both Chinese and Pakistan sides with equal consultation, full analyses, and joint implementation. Pakistani environmental laws and regulations will be strictly abided by the implementation of relevant projects. For these projects, the responsible parties need to propose an environmental impact assessment report and make sure the environmental performance of projects during implementation will meet or surpass national environmental quality standards. In planning the future cooperative projects, both sides will pay more attention to the environment and people's livelihood and give more considerations to clean energy projects like hydropower and natural gas.

(2) Suggestions on green development on the CPEC

During constructing CPEC, it is necessary to incorporate environmental considerations into the overall plan as early as possible. For example, while attempting to reduce transportation costs, the roads, railways, and pipeline construction of CPEC might cause damages to the natural ecosystem and cause biodiversity loss. Other construction projects such as constructing hydropower stations could also cause fragmented habitats, forest degradation, groundwater pollution, and soil pollution.

In February 2016, the Parliament of Pakistan included SDGs into national development goals, which were further incorporated into *Pakistan Vision 2025* subsequently. Pakistan adopted National SDG Framework, and clarified key goals, index, and baseline. The Belt and Road Initiative and green development of CPEC are directly related to the implementation of SDG 9, i.e. "promote inclusive and sustainable industrialization and, by 2030, significantly raise industry's share of employment and gross domestic product, in line with national circumstances, and double its share in the least developed countries." In addition, the Belt and Road Initiative and green development on CPEC will also provide support for realizing other specific targets of SDGs.

Firstly, the Chinese and Pakistani government could carry out joint actions to conduct a strategic environmental assessment on CPEC. The Chinese and Pakistani governments should make sure all the EIA of CPEC projects meet certain standards and guarantee the implementation of relevant measures. Secondly, they need to focus on key biodiversity regions and adopt special protection measures targeted at these regions. For example, IUCN is collaborating with the Pakistani government and carried out projects in the northern part of Pakistan. This project aims to study and cope with the influence of CPEC on the biodiversity of this region. Lastly, it is important to promote the active participation of the public, exploring crowdfunding to support the restoration of ecosystem, afforestation from farmland, and carbon emission cut.

4.1.2 The ecological and environmental protection of the Belt and Road projects in Sri Lanka

Sri Lanka is known as the "Crossroad of the Indian Ocean". The Hambantota

Harbor in the south is only 10 n mile away from the main channel of the Indian Ocean. It is a place that must be passed on the shipping line that connects the Middle East, Europe, Africa, and the East Asian continent. It is also a key node of the 21st Century Maritime Silk Road.

Since the implementation of the Belt and Road Initiative, China and Sri Lanka carried out pragmatic cooperation on harbor, power plants, telecommunications, water supply, water conservancy, and highways. In general, the projects are selected based on the urgent need for economic and social development in the local area and scientific analyses of previous governments. Norochcholai Power Plant provides about 40% of national power, making frequent power outages in Sri Lanka now a history. The Colombo Airport Expressway brings convenience to tourists and promotes the rapid development of Sri Lanka's tourism industry. The Moragahakanda Dam is a large-scale water conservancy project invested and contracted for construction by Chinese enterprises, which provides great benefits for local agriculture and people's livelihood. In 2017, China and Sri Lanka signed the *Framework Agreement on Promoting Investment and Economic and Technological Cooperation*. The Ministry of Commerce of China and the Ministry of Development Strategies and International Trade of Sri Lanka signed *the Mid and Long-term Planning Outline of China and Sri Lanka on Investment and Economic and Technological Cooperation*. Against this backdrop, a series of transportation infrastructure projects are implemented steadily with Port City of Colombo, Hambantota Harbor operation, and the industrial park at Hambantota Harbor among other major cooperation projects actively facilitated. China-Sri Lanka cooperation created over 100,000 jobs and trained tens of thousands of technological and management talents.

According to the *National Environmental Law* and *Environmental Impact Assessment Law* of Sri Lanka, there are strict EIA procedures in Sri Lanka. The project implementers need to formulate EIA report and their projects could be inaugurated only if the EIA reports are approved by responsible government branches on environmental governance, such as the Central Environmental Authority, Coast Conservation, and Coastal Resources Management Department and administrations on marine resource management. The Belt and Road projects with the participation of Chinese companies could strictly abide by relevant environmental protection policies and institutions of Sri Lanka in general.

4.1.3 Major Enlightenment for Greening BRI

Principles applying to all BRI projects more generally follow:

(1) Aligning BRI and *2030 Agenda*

- The objectives of the BRI closely align with the spirit of the *2030 Agenda* and *Paris Agreement* on Climate Change. The concept that cumulatively BRI projects have to stay within the limit of the planet would be useful to state.
- There is an intended shift from the first BRI phase oriented to mega infrastructure projects to a second phase BRI oriented to the establishment of special economic zones, social investments, and a greater focus on environmental issues.
- China is today the champion of environmental multilateralism and green leadership (Paris climate agreement, host of CBD COP15, fight against pollution etc.).
- All new BRI-related organizations and networks being created (BRI Green Coalition, BRI Green Cooling Initiative, BRI Green Lighting Initiative, BRI Green Going-Out Initiative) need to reinforce existing multilateral institutions and the *2030 Agenda*.
- Green finance and investment and CSR standards need to be implemented by Chinese investors and project implementation companies. The BRI Green Investment Principles, supported by 27 international banks and investors, and BRI Green is a good start. Via its new Going-Out Initiative China could export its green finance norms and standards.

- Strong convergence is needed between the principles of implementation of the *2030 Agenda* and the BRI implementation agenda on national trajectories for transformation towards sustainable development, which are country-specific.

- The alignment between the BRI objectives and *2030 Agenda* is not reflected in the BRI energy projects which are targeting mostly fossil fuel projects while the NDCs of BRI countries indicate strong need for investment in renewables.

(2) Enacting principles that ensure new projects are green from the start

- There is a need for principles that ensure new projects are green from the start – these include creating a more environmental-focused mind set amongst donors; developing guidelines for development banks in China to think long-term, building in the environmental factors; and starting pilot projects within a 5-year time horizon where infrastructure designs is aligned with ecological concepts (e.g. nature-based solutions for infrastructure).

- Holistic and integrated approaches need to be used to combine the dynamics of multiple projects and evaluate their cumulative impact. Ecological approaches such as the one behind the development of the Yangtze River Economic Corridor, holistic in nature, cost-effective and bringing extensive ecological benefits can be followed.

- Special Economic Zones (SEZs) need to also be Environmental SEZs. Benchmarks for emission-free economic zones need to be in place.

- Ex ante and ex post evaluation by Chinese-international teams including academics and think-tank experts is needed of the environmental and social impacts and alignment with *2030 Agenda* and its implementation in host countries.

- Facilitation of South-South cooperation addressing the main environmental issues in BRI is needed.

At each BRI project level:

- As a minimum: avoiding, reducing and compensating local environmental and social damage (by providing environmental and social safeguards). It involves stakeholder consultation and participation.
- Evaluating the environmental impact of the project using internationally established methodologies integrating long-term impacts and potential irreversibility.
- Evaluating the alignment of the project with the implementation of the *2030 agenda* and the *Paris Agreement* in the host country. Avoiding path dependencies and lock-in effects of fossil-fuel investments and non-resilient infrastructure.
- Address the challenge of complexity and sometimes transnational nature of projects.

At the Corridor/program level:

- Ensuring environmental sustainability, cumulative impact evaluation and policy coherence;
- Understand how the Corridor/program fits within the structural transformation of the country towards low carbon economy and national debt level.

At whole BRI level:

- Compatibility of the BRI impact on global production, trade and exchanges intensification with the global CO_2 reduction needed by 2050.

(3) Understanding the role of the Digital BRI can play

- There is lack of studies on the BRI and its impacts developed by host country research institutions due to lack of access to project data and sometimes low capacity and expertise.
- The Digital BRI, platform for participating countries to share the data obtained as part of their collaborative projects with each other and with China, and particularly the BRI Environmental Big Data Platform launched

at the Second BRI Forum and aimed at sharing environmental performance data of BRI projects can help. However, such developments also have implications regarding China's rising position as a science-development superpower with tens of thousands of researchers and students, and hundreds of universities in low- and middle-income countries involved.

- The creation of a fund for BRI Studies Network, which would finance independent research consortia, recognized by the various BRI stakeholders, could be a possible solution.
- The Digital BRI could facilitate an exchange of methodologies practiced by each country.

(4) Implementing projects which are demand-driven and sustainable

- Eagerness of developing countries to engage with power economies like China can overlook local needs for sustainable development and translation of the *2030 Agenda*.
- Host countries can have an important gap between high environmental aspirations in relations to development projects and their implementation.
- There is a significant gap in some host countries in understanding regarding how to deliver the same.
- There is a need to raise awareness that economic benefit of investing in green solutions based on local context and native design and landscapes can be the same or even greater than conventional investments.
- Appropriate channels and signals need to be in place to engage with stakeholders (local or independent research organizations, local industry, civil society organisations) on social, environmental and economic concerns related to the BRI so that their inputs are valued and taken into account. In Pakistan, some Chinese companies are working closely with local and international NGOs, and this provides a model to be expanded.
- Successful and environmentally-sound implementation of BRI projects in host countries such as SL and PK could play an demonstration role to

projects in other participating countries.

(5) Engaging in concrete environment-related projects

- Identify high priority, well-defined projects which are workable and clearly show environmental benefits.
- Engage in concrete cooperation, bring Chinese best practices and provide technical and/or financial assistance on environment-related issues possibly via twinning projects. Examples include floodwater management, water governance, ecological redlining, National Parks management, large-scale forest restoration, restoration in arid zones, etc.

4.2 China-Malaysia Qinzhou industrial park

China-Malaysia Industrial Park located in Qinzhou, Guangxi Zhuang Autonomous Region, is one of the most exemplary cases on BRI green development. Guangxi plays an important role in the development of the Belt and Road Initiative with special advantage in green development. Located in the South of Guangxi, Qinzhou is at the intersection of the Silk Road Economic Belt and the Maritime Silk Road and the frontier of China-ASEAN cooperation. In recent years, Qinzhou has been attaching great importance to ecological civilization construction and green development, especially the green development of China-Malaysia Qinzhou Industrial Park (CMQIP). It is believed that the experience of CMQIP in green concept promotion, system construction, planning and policy implementation is of great value to other countries and regions in greening the Belt and Road.

CMQIP is a flagship project of China-Malaysia investment cooperation. With a planned area of 55 km^2, CMQIP prioritizes the development of bio-technology, medicine, electronic information, equipment manufacturing, new energy and new material, modern services and major industries of ASEAN. CMQIP aims to build CMQIP into a high-end industrial cluster, a model park for industry-city integration, an area with rich scientific, education and human resources, a pilot zone of international cooperation and free trade as well as a flagship project of China-Malaysia cooperation and a role model zone of China-ASEAN cooperation. CMQIP Phase I covers an area of 15 km^2, including the 7.87 km^2 start-up area. The positioning of CMQIP is an advanced manufacturing base, an information corridor, a modern town of culture and a platform for cooperation & exchange. China-Malaysia Qinzhou Industrial Park is positioned as: 1) advanced production base, with priority on cultivating emerging industries that require advanced production like biology and multimedia. 2) Information smart corridor. It will draw on the successful experience of Multimedia Super Corridor (MSC) of Malaysia to develop an industry-education-research integrated smart park on biology and multimedia. 3) A new city of culture and ecology. It will organically combine the urban space of China-Malaysia Qinzhou Industrial Park, natural mountains in its neighboring areas, and major river systems (Jingu River), to create an artistic space featuring green ecology with the combination of mountains, rivers, and cities. 4) The window for cooperation and exchange. With its advantageous geographical location to ASEAN countries, China-Malaysia Qinzhou Industrial Park will be developed into an information release platform, trade exchange platform, and window for project showcase and business cooperation for China-ASEAN Free Trade Area.

4.2.1 Practices of CMQIP in promoting green development

(1) Incorporating green development in China-Malaysia Qinzhou Industrial Park Regulation

Legislation for industrial parks is rare in China. However, to support and promote the development of China-Malaysia Qinzhou Industrial Park, the Standing Committee of the 12th People's Congress of Guangxi Zhuang Autonomous Region adopted *China-Malaysia Qinzhou Industrial Park Regulation* in 2017. The *Regulation* is the bases for the development of CMQIP, content related to green development include:

Article 28 of Chapter 4 has stipulates: CMQIP shall follow the concept of green development, build a green industrial system and spatial layout, improve ecological protection facilities and measures, promote eco-industries, establish an energy-conservation, environment-friendly and smart industrial park, and build a new eco-city which integrates industries and urban areas.

Article 34 of Chapter 4 stipulates: CMQIP shall establish and improve its system of ecological and environmental protection indicators; promote recycle of energy, comprehensive use of water resources, reduction, innocuous treatment and recycle of waste; and control the total discharge of key pollutants and promote low-carbon and circular economy.

CMQIP shall toughen environment threshold on market access; improve construction of environment infrastructure; promote the environment industry; put a ban on industrial projects with high energy consumption, high pollution and high environmental risks; support the development of enterprises with low energy consumption and low emission; encourage clean production in businesses; and protect and improve the environment.

In light of this, China-Malaysia Qinzhou Industrial Park highlights green development with strict industry access and prohibits developing industries with high energy consumption, high pollution, and high emission.

(2) Incorporating green development in the Master Plan of CMQIP

Master Plan of CMQIP was approved by the government of Guangxi Zhuang Autonomous Region in June 2013. The Master Plan identified the six overall requirements of high-quality development, industry-city integration, innovation in mechanism, openness and win-win cooperation, wealth and harmony, green development, in which "green development" is one of the major goals and principles.

The Master Plan was amended in January 2018 which specifies and standardizes the conception of green development for the industry and the park, which provides solid planning guidance on the green development of the industrial park. The Master Plan highlighted green industrial development through adjusting the six priority industries of equipment manufacturing, electronic information, food processing, traditional and new materials, bio-technology and modern services to seven priority industries of electronic information, smart manufacturing, bio-medicine, new energy, new materials, modern services and traditional ASEAN industries. Equipment manufacturing with high energy consumption and food processing with heavy pollution are removed from the list of priority industries.

In addition, the detailed development plans of the Qinzhou Park further underscored the construction goal and principles of "integrated overall planning, people-oriented, creating characteristics, and low-carbon and harmonious," emphasized the adoption of clean production technologies and exploration of constructing ecological chain and ecological network within the industrial park, maximize resource use efficiency and valued the protection and construction of natural landscape in the industrial park.

(3) CMQIP is committed to building itself into a national green industrial park

Related departments of the central government (e.g. Ministry of Industry and Information Technology) attach great importance to the development of green manufacturing and green industrial parks and have released *the Green Industrial Development Plan (2016—2020)*, *Guidelines for the Development of Green Manufacturing Standards* and *Industrial Energy Conservation and Green Standardization Action Plan (2017—2019)* to encourage and support the green development of industrial parks. As an industrial park under construction, CMQIP has been regarding becoming a national green industrial park as its primary goal; CMQIP has been promoting and improving the construction and management of the Park according to related documents released by the central government.

(4) Implementing ecological redlining

CMQIP Authority commissioned professional agencies to work out *Research Report on Ecological Redlining of China-Malaysia Qinzhou Industrial Park*. The Research report identified the ecological redline of CMQIP, defined Class 1 ecologically controlled area that includes the mangroves and Class 2 ecologically controlled area that includes wetlands. The Research Report also identified specific rules and regulations for the two kinds of ecologically controlled areas. Damaging and forced restoration of mangroves are prohibited in Class 1 ecologically controlled area; development activities are restricted and ecological conservation and restoration efforts are encouraged in Class 2 ecologically controlled area. The Research Report also identified the proportion and layout of space for ecological functions.

(5) The municipal government of Qinzhou incorporated green development into the assessment of CMQIP

The municipal government of Qinzhou carried out assessment of CMQIP according to *Implementation Plan for the Trial Assessment of Industrial Parks in Qinzhou*, covering water and soil conservation, geographical disaster risks, mineral resources, earthquake resistance capacity, climate impact and the protection of cultural relics[①].

4.2.2 Initial progress of CMQIP in green development

First, environmental quality in CMQIP remained stable. CMQIP attaches great importance to the development of distributive energy and solar power generation. Data captured by local environmental authorities show no significant difference in environmental quality in CMQIP and other areas of Qinzhou City. Besides, the water environment functioning zone met water quality standards with no significant water pollution incident.

Second, CMQIP witnessed continual improvement in major environmental indicators. CMQIP achieved a better performance than the national average of green industrial parks in major indicators for green development, including per unit GDP energy consumption, water consumption, COD, SO_2 emission, NO_x emission, ammonia nitrogen emission and the comprehensive utilization rate of industrial solid wastes.

Third, progress has been made in the development of green buffer zones. The

---

① The municipal government of Qinzhou. The municipal government of Qinzhou carried out assessment of CMQIP according to Implementation Plan for the Trial Assessment of Industrial Parks in Qinzhou[EB/OL]. 2019-01-08. http: //www. qinzhou. gov.cn/phoneSite/zcjd/201901/t20190108_1977888.html.

buffer zones on both sides of the road and the coast have been basically built to effectively protect mangroves and wetlands.

Fourth, industries developed in CMQIP meet the requirements for green and environmental-friendly industries. CMQIP developed and implemented strict rules for the approval of companies to be settled in the Park according to the Master Plan. By the end of 2018, 350 companies have settled in CMQIP. They are mainly engaged in modern services, including financial services, modern logistics and cultural development, modern manufacturing, including electronic information and modern equipment manufacturing, and strategic emerging industries, including bio-medicine, Nano-technology and cloud computing.

4.2.3 Experience of the green development of CMQIP

First, sticking to the concept of ecological priority and green development. The Chinese government is promoting ecological civilization construction nationwide with ecological priority and green development as the core. Ecological priority means that development should never cross the ecological red line and should always regard ecological conservation as an important goal; green development means promoting energy conservation, environmental protection and ecological conservation.

Second, incorporating green development into every aspect of Park planning. Green development must be incorporated into the plans for the development, operation and management of the Park to ensure that all activities in the industrial park and all functional zones follow the principles of green development. Therefore, it is necessary to incorporate green development concepts, including energy conservation, environmental protection and ecological conservation, into the construction and development plans of the industrial park.

Third, incorporating green development into legislation. The construction and development of industrial parks is a long-term, complex process that requires the support of legislation. To ensure the implementation of green development concepts in the construction and operation of industrial parks, it is necessary to incorporate green development into industrial park legislation, making compliance with green development principles a legislative duty.

Fourth, incorporating environmental considerations into the approval of projects. Industrial parks are established to promote industrial development. To ensure green development of the parks, projects with high energy consumption, heavy pollution and occupation of ecological resources should be prohibited. An inspection and approval system needs to be established to ensure that only green projects are allowed in the park.

Fifth, establishing a green development dynamic assessment mechanism for industrial parks. It is necessary to establish a dynamic assessment mechanism for the green development of industrial parks to identify projects and businesses failing to comply with green development principles. Related projects and businesses should be asked to solve the problem within a certain period of time and should be expelled from the industrial park in case they failed to do so. Therefore, a project withdrawal mechanism should also be established.

5 Policy Suggestions on Promoting Green Belt and Road

The United Nation *2030 Agenda for Sustainable Development* is a far-reaching framework for all the countries in the world as the goal and consensus to develop and achieve in the future. However, the significance and benefits of building green Belt and Road may differ for countries along the route. Only when all the parties recognize the positive role of greening Belt and Road in their long-term sustainable development can cooperation under the Initiative advance. Therefore, firstly strategic coordination between countries' implementation plans of sustainable development and the green Belt and Road Initiative, should be enhanced without violating the main principles of the Belt and Road Initiative, including inclusiveness, coordination, consistency and capacity-building(Figure 5-1). Secondly, by understanding the essence of

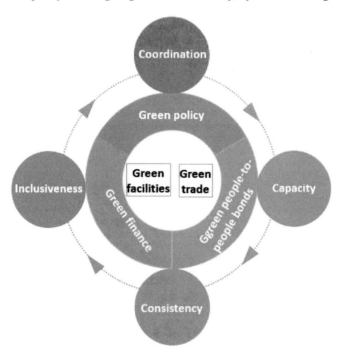

Figure 5-1 Roadmap for the development of a Green BRI

the green Belt and Road Initiative, the concept of green development needs to be embedded into the efforts in achieving the "Five Goals" (namely policy coordination, facilities connectivity, unimpeded trade, financial integration and people-to-people bonds) and reduce the adverse effect to ecological environment during the implementation of the Belt and Road Initiative. Greenization of activities related to promoting "facilities connectivity" and "unimpeded trade" is the focal point while the goals of "green policy coordination", "green financial integration" and "green people-to-people bonds" could function as policy and financial support and be conducive to create an amicable external atmosphere.

5.1 Play an active role in global environmental governance and climate governance, transforming the Belt and Road Initiative into an important instrument for global ecological civilization construction and building a green community of common destiny

Forging the international partnership and network for green development on the platform of green Belt and Road. The international community needs to make concerted efforts to respect nature and to stress green development. Ecological civilization construction is not only an effective attempt of China in promoting sustainable development as the largest developing country in the world, but also an important concept and path provided by China for global environmental governance. In developing the Belt and Road, the concept of ecological civilization to provide Chinese wisdom to constructing green human community of shared destiny as well as models and experience that could help late-comers in green development to avoid reliance on traditional path and lock-in effect need to be followed so as to help more countries and regions to accept and implement sustainable development.

Enhancing ecological and environmental cooperation on the Belt and Road through establishing the BRI International Green Development Coalition. The Coalition is an open, inclusive and voluntary international network. The Coalition will provides a platform of policy dialogue and communication, sharing ecological civilization and green development, and facilitating green development into the implementation of BRI. It will promote international consensus and collective actions. Meanwhile, the Coalition will provides a platform of knowledge and information, bringing together international think tanks, carrying out joint research and jointly promoting Belt and Road participating countries to implement the *2030 Agenda*.

Participating in global environmental governance and jointly promoting ecological civilization construction worldwide on the platform of green Belt and Road. Active involvement in the reform and construction of the global environmental governance system is needed. Moreover, an increase in the awareness of countries along the Belt and Road in building a community of shared destiny with shared responsibilities and benefits and develop international rules through joint discussion and negotiation to conduct global environmental governance is also needed. Launching fair rules for global environmental governance with the interest of all sides being taken into consideration, promoting capacity building in global environmental governance, offering a new path and new plans for the implementation of the *2030 Agenda* on a global level and advancing the global environmental governance system could be elements of solution.

Promoting the concept of ecological civilization and facilitate consensus on ecological civilization achieved on the platform of green Belt and Road. The concepts of ecological civilization and sustainable development are both formed against the background of deep adjustment in the global economic governance system. The two concepts aim at building a green home for the human society, and therefore have a lot in common regardless of their difference in perspectives, stands and expression of the final goal. As a result, promoting the alignment of the development of the green Belt and Road and the United Nations *2030 Agenda* and the mutual learning, understanding and support of ecological civilization in China and sustainable development in countries along the Belt and Road could advance the global green agenda.

5.2 Promoting strategic alignment in the development of the green Belt and Road with connection of policies, planning, standards and technologies

5.2.1 Promoting strategic alignment in constructing a green Belt and Road

Integrating green development into the Belt and Road projects as an important part of the MOUs on BRI construction between China and related countries and international organizations. Any MOU should include contents related to ecological civilization and green development concepts, jointly building a green Belt and Road, promoting the alignment of the Belt and Road development with the United Nations 2030 Sustainable Development Goals (SDGs). Making full use of the bilateral strategic agreements between China and countries along the Belt and Road and international organizations to establish ecological and environmental work groups who are responsible for the alignment of strategies and plans on green development is needed.

Making full use of platforms for communication to promote strategic alignment. Negotiation with related parties is needed to establish a fixed panel on the development of green Belt and Road under dialogue mechanisms such as the Forum on China-ASEAN Environmental Cooperation, Euro-Asia Economic Forum and China-Arab States Environmental Protection and Cooperation Forum. It is also suggested to set up a parallel panel on green development of the Belt and Road. The Panel aims to discuss project construction, financing guidelines and technical standards for the development of the green Belt and Road to promote strategic alignment in standards and rules. On the other hand, it is important to establish a standing work group on the ecological environment with partners based on the bilateral agreement on strategic cooperation signed with BRI participating countries and international organizations. The work group will be specially responsible for the connection of green development strategies and planning of both sides. It's necessary to list cooperative projects on green Belt and Road as a priority area for the future work of bilateral and multilateral cooperation mechanisms, including China-Russia, China-Kazakhstan, China-Europe, China-ASEAN, Shanghai Cooperation Organization and BRICS to facilitate strategic connection with cooperation on specific projects.

Promoting ecological and environmental protection policy alignment with hosting countries. BRI projects need to be strictly monitored so to have a clear understanding of their impact on local and global environment; the implementation of an international partnership can support a successful monitoring system. The partnership should be involved in outling a BRI policy coordination and in defining an overall and internationally accepted notion of "Green vs. Non Green".

Promoting alignment of the Belt and Road Initiative and South-South Cooperation. The Belt and Road Initiative overlaps with South-South Cooperation in many aspects. Therefore, strengthened efforts in promoting the Belt and Road strategy under the framework of South-South cooperation to make it a model for South-South cooperation between China and developing countries is needed. An international research group needs to be established to help developing countries to better understand green development and sustainable development.

5.2.2 Integrating ecological and environmental cooperation into the whole process of implementing BRI

Enriching and stressing content related to ecological and environmental cooperation in the planning for the development of the Belt and Road Initiative. It is necessary to formulate the *Guidelines for Compiling Belt and Road Development Cooperation Plan* that mandate the inclusion of ecological and environmental protection and cooperation on green development in Belt and Road development planning and guide authorities to enrich the content of related chapters in comprehensive plans and sector-specific plans and include the building of green Belt and Road as a key part in related documents in the future.

Jointly working on ecological and environmental protection plans with the BRI countries, particularly those with multiple collaborative projects. Joint analysis of ecological and environmental protection and green development plans with BRI countries together with infrastructure connectivity and international industrial capacity cooperation plans and prominent ecological and environmental challenges is needed. Such analysis needs to be updated regularly.

Facilitating cooperation and application of environmental standards. Efforts on the coordination of standards for green infrastructure of China and Belt and Road countries are needed. Through joint research, a set of international standards widely acknowledged by Belt and Road countries, in green transportation, green architecture, and green energy should be developed. Relying on the Belt and Road Center for Environmental Technology Innovation and Transfer, as well as on the demonstrative base for industrial cooperation, support to businesses cooperating within Belt and Road countries, and jointly releasing environmental industry standards recognized along with relevant associations should be provided.

Promoting technical alignment among different BRI projects. BRI projects are mostly transnational and extremely complex to develop due to stringent technical constraints. A special commission should be established in order to develop, over the years, agreed major BRI technical standards, defined for key sectors, with the intent to simplify public tenders and international bidding procedures.

5.3 Safeguard mechanisms for constructing a green Belt and Road from its source, and guiding green investment with mechanisms of green finance and ecological impact assessment

5.3.1 Supporting the development of Green Belt and Road with green finance

Adopting green financial tools for greening the Belt and Road on the global level. First, Belt and Road green investment and financing principles need to be further studied. Principles and guidelines for Belt and Road green investment and financing according to the United Nations 2030 SDGs and the Paris Agreement in reference to international rules and standards need to be developed. These principles and guidelines, once launched, need to be promoted in countries along the Belt and Road. Second, multilateral Belt and Road green investment and financing guarantee institutions need to be established to provide guarantee for investment and financing in green projects and energy conservation and emission reduction projects to share risks and attract the private sector to invest in green industries. The guarantee institutions are non-profit, market-oriented institutions guiding and promoting the development of Belt and Road projects. Third, an environmental and social information database on the Belt and Road Ecological and Environmental Protection Big Data Service Platform needs to be established to provide information services to investors, lenders, financiers and property owners along the Belt and Road. Fourth, financial institutions along the Belt and Road need to disclose environmental information, so as to promote the green development of financial institutions while helping businesses to pursue green development and achieve green economic development and improvement in environmental quality.

Encouraging countries along the Belt and Road to regard green finance as an important instrument to realize the transition to a green development model. Promoting the capability of BRI countries in developing green finance and encouraging them to share experience in this field. First, by encouraging the development of green industries, sectors and customers and fostering growing demand for green investment and financing. Second, by launching financial regulatory policies to encourage financial institutions to actively support the development of green industries, sectors, companies and customers, guide and encourage financial institutions to establish green investment and financing mechanisms and form a long-term mechanism for promoting green economic and social development with green finance. Third, by nurturing environmental and social responsible investors.

Giving full play to the role of financial institutions as agents to improve the

environmental performance of corporate customers. First, by encouraging financial institutions to establish a clear strategy for the development of green finance. They need to identify the strategic goals for green finance, foster green philosophy and values, improve the organizational structure of green finance and actively expand the market of green finance, so as to effectively control environmental and climate risks. Second, by establishing and improving green finance policies and systems for overseas operations. Environmental and climate factors present both risks and opportunities, which require financial institutions to carry out in-depth research to assess and effectively control environmental risks and launch innovative green finance products for optimized green finance services. Third, by establishing an environmental and social risk assessment toolkit on the basis of international standards such as IFC Environmental and Social Performance Standards and the Equator Principles, to analyze the environmental and social conditions along the Belt and Road. Fourth, by ensuring the lifecycle management of environmental and social risks through incorporating environmental and social risk management into credit and investment management before, during and after an investment is made. Fifth, by implementing environmental information disclosure through establishing the system and framework of environmental information disclosure and improving the capability of information disclosure. Sixth, by designing a mechanism to address environmental and social risks.

5.3.2 Establishing ecological and environmental management mechanism for Belt and Road projects

Establishing the environmental management rating mechanism for Belt and Road projects. An environmental management database for Belt and Road development and investment projects needs to be set up to include ecological and environmental impact into the Belt and Road development project risk rating system and conduct risk assessment to Belt and Road development projects in terms of economic risks, political risks, social risks, cultural risks and ecological and environmental risks. Identification of the ecological and environmental impact of investment projects on different dimensions such as ecological security and environmental pollution and assess the environmental benefit of these projects is also needed. The assessment results could become an important reference for developmental and policy-based financial support.

Developing environmental management tools for Belt and Road projects. Tools for the identification, assessment, monitoring and management of environmental, climate and social risks of development and investment projects, carry out research on investment and consulting service tools with full visibility to policy, laws and regulations, data and information and develop country-specific ecological and need to be developed along with environmental information system and assessment tools for key investment destinations along the Belt and Road as well as technical supporting tools for greater availability of public environmental data. Comprehensive and in-depth assessments of ecological and environmental risks in ecologically sensitive and vulnerable areas need to be conducted along with establishing lists of risks and control measures and requiring investment projects to carry out the tasks and requirements of related lists. In terms of the range of assessment, the inclusion of ecological health and climate change into environmental assessment on a voluntary basis with regard to international hotspot issues needs to be promoted.

Improving the environmental management platform and procedures for Belt and Road projects. An environmental protection assessment consulting service platform along with an investment and financing information sharing platform among governments, businesses, banks and other stakeholders are needed to conduct independent assessment and discussion of "Belt and Road" development project information disclosure and to ensure the implementation of environmental and social security measures of related projects and protect the interests of stakeholders.

Encouraging the participation of stakeholders in the environmental management procedures. Stakeholders need to be encouraged to comprehensively and effectively engage in environmental impact assessment and implement projects strictly in line with legislations and regulations in the hosting countries. It is important to ensure scientific judgement of environmental impacts.

5.4 Constructing a mechanism for BRI project management and promoting the business to adopt practices on green development

Strengthening green supply chain management. Integration of China's and international competitive industries into a green global supply chain system is needed. By joining government departments, institutions and enterprises of the Belt and Road countries together, a regional green supply chain system can be developed. Each country can bring its industrial and market advantages into full play, which makes international cooperation more extensive, deeper and to higher level. Besides, full use of the Belt and Road green supply chain cooperation platform to support and encourage enterprises should be made to actively get involved in foreign trade and investment cooperation and to promote green innovation. Pilot demonstration projects of green supply chain management, development of green supply chain environmental management policy tools and promotion of green development of the entire industrial chain of production, distribution and consumption should be carried out. A green supply chain performance measurement indication system to evaluate the performance of the enterprises and improve corporate social responsibility for sustainable development needs to be developed.

Exploring to set up the Belt and Road Green Development Fund. It is necessary to enhance financial input, and safeguard the implementation of green Belt and Road related work. We need to promote to set up a special fund for resource development and environmental protection, with priority given to supporting environmental infrastructure, capacity building and green industries of countries along the Belt and Road. Meanwhile, it is necessary to give full play to the leading role of policy banks such as China Development Bank and the Export-Import Bank of

China, to guide and encourage the pooling of all kinds of resources to the Green Development Fund. It will support and bring new momentum to green Belt and Road.

Developing green value chain. First, China should strengthen the capture and promotion of the spillover of green technologies of developed countries, improve the capability of developing countries in adopting the green technologies of developed countries and promote the application of the advanced green technologies in BRI participating countries along the Belt and Road. Second, China should promote the development of Green Value Chain in international productive capacity cooperation and industrial parks through promoting the recognition and application of industrial rules and regulations on environment to promote energy conservation and emission reduction in industries with heavy pollution, and developing incentives and supportive policies to nurture and develop clean energy and other environmental-friendly industries. On the other hand, it is imperative to enhance support to clean energies like nuclear power, hydropower, wind power, and solar power, etc., as well as other environmentally- friendly industries, safeguard the cultivation and implementation of projects and facilitate the sustainable and coordinated development of economy, society, and environment. Third, China should guide the development of Green Value Chain with green standards and labeling. It is necessary to develop green standards for key stages of the product lifecycle and players on the value chain, ensure environmental-friendly input and output along the value chain and establish green standards and assessment and certification mechanisms for raw materials used in production and the finished products.

Facilitating trade in environmental goods and services. Facilitating trade in environmental goods and services will bring huge environmental benefits to BRI participating countries along the Belt and Road. Increasing market openness for environmental products and services, developing green industries, and encouraging the import and export of environmental products and services related to air pollution control, water pollution prevention and control, as well as solid waste management and disposal technologies and services to be promoted. Differentiating policies for green products and non-green products, and providing classified guidance and management for product trade and investment, such as reducing tariffs of green products, offering special or preferential customs treatment, listing green products as an encouraged category for industrial investment, as well as providing green financial services is recommended. Seeking concessional loans for green projects from international financial institutions such as the World Bank and the Asian Development Bank is also encouraged. Besides, improving green product identification and labeling systems and mutual recognition of green products between countries, as well as encouraging more procurement of products with eco-labels by governments along the Belt and Road is also recommended.

5.5 Building a green Belt and Road through enhancing people-to-people bond, and enhancing personnel exchange and capacity building

Making the Green Envoys Program for Maritime Silk Road the flagship activity for environmental protection capacity building. The Green Envoys Program (upgraded to Green Envoys Program for Maritime Silk Road in 2016) functions as an important platform for China to carry out South-South environmental cooperation and to promote regional sustainable and green development. The Green Envoys Program for Maritime Silk Road should be the flagship activity, capable of enhancing public awareness on environmental protection and strengthening capacity building in the BRI participating countries. By advancing "policy communication" and facilitating "people-to-people bonds", China will strengthen cooperation and exchanges in environmental management, pollution prevention and control, green economy and other fields through providing environmental management personnel and professional training, as well as policy guidance, and share China's ideas and practices realizing ecological civilization and green development. Local governments should be encouraged to participate in the Green Envoys Program for Maritime Silk Road, and guide environmental protection enterprises to "go global" in an orderly manner via platforms such as China-ASEAN Demonstration Base for Environmental Technology and Industry Cooperation and the Belt and Road Environmental Technology Exchange and Transfer Center (Shenzhen).

Supporting and promoting exchanges and cooperation of environmental protection social organizations between China and countries along the route. A supporting network ensuring government guidance, enterprise support, social participation, and industry mutual assistance needs to be established. Roles of government entities that are responsible for cooperation should be clarified and overseas environmental responsibilities of Chinese enterprises should be defined by issuing policies or guidelines to attract various parties and involve and encourage environmental social organizations to establish a cooperation network of their own. Diversified financial mechanisms and increase government procurement of services by environment protection organizations should be formulated. Special cooperation funds to support environmental protection social organizations to go global should be set up. Participation mechanisms for environmental and social organizations should be designed so to encourage their involvement in negotiations and decision-making, and create an international communication event list for environmental and social organizations.

Promoting gender mainstreaming and improving the leadership of women. It is necessary to improve the awareness of policy makers on the role of women in social and environmental development and promote the integration of gender mainstreaming in Belt and Road policy development and project implementation. It is important to implement the best practice of gender mainstreaming in the development of Belt and Road projects and promote female government leaders, experts and young scholars in the ecological and environmental protection sector in countries along the Belt and Road to engage in the training program of "Improving the Green Leadership of Women" and share with Belt and Road partners the methods and experience of realizing gender mainstreaming with the help of Green Envoys Program.